TESLA COIL SECRETS

Construction Notes and Novel Uses

R. A. Ford

Tesla Coil Secrets

Construction Notes and Novel Uses

by R. A. Ford
Simplified Technology Services, Urbana, IL

ISBN 0-917914-31-7

5 6 7 8 9 0

CONTENTS

Notice of Disclaimer

The author of this book is not a professional engineer nor has he had formal training in the design or operation of Tesla coils. The methods that he describes are presented merely as guidelines for others in developing Tesla coils.

The construction of Tesla coils can be very dangerous, and dangers have been pointed out wherever possible. Since the author is not a professional in this field, there are probably other dangers as well.

Neither the author nor Lindsay Publications, Inc. has built the Tesla coils described. Both hereby disclaim any liability for injury to persons or property that may result while using this book. Both do not intend by this publication to explain all dangers known or unknown that may exist in the building and operation of the projects described.

Write for a catalog of other unusual books available from:

Lindsay Publications, Inc.
P.O. Box 12
Bradley, IL 60915-0012

"The whole world here unlocks the experience of the past to the builders of the future."

(Inscription over the east entrance, main library, University of Illinois, Urbana)

The author appreciates resources made available by the library staff at the University of Illinois, Urbana, without which this book would not be possible.

Part One - General Introduction

The Tesla coil, invented by Nikola Tesla in the early 1890's consists of two coaxial, concentric single layer coils wound in the manner of a step-up transformer, each coil having an air core. The coil unit produces high voltage, high frequency electricity.

The Tesla coil transformers came into being as equipment developed by Nikola Tesla to explore the nature of tuned circuits resonating at high frequency and high voltage. He discovered early in his researches that while using a coil of a given wavelength, other coils in the laboratory, tuned either to the same wavelength or one of its harmonics, would respond in sympathy by spouting its own crown of sparks, even though not connected in any physical way to the operating coil.

Here was an example of transmission of electric power over distance without wires. Unlike shortwave Hertzian wave generators, Tesla's method was far more efficient in transmitting usable amounts of power because he was experimenting at much lower frequencies which had great penetrating ability.

Tesla's wireless power transmission system began with "revelations" he received while observing a great thunderstorm on July 3, 1899, at Colorado Springs. He concluded that the earth, being a conducting body of fixed dimensions insulated in space, should behave as though it were a spherical electrical condenser. He then began his now famous Colorado Springs experiments to periodically charge the earth and test his theory.

Up to this time, power was supplied by such stations as the hydro-electric system at Niagara Falls using Tesla's polyphase generators. Even though Tesla had masterminded the complete system, transmission of power via wires to distant places required high voltage to reduce large line losses.

Tesla claimed his wireless power transmission system to be about 98% efficient but industrial forces gaining momentum with the Tesla polyphase system could not be expected to be interested in new experiments which would destabilize the existing method of electrical power distribution.

Wireless power transmission was only one part of an overall plan Tesla had conceived in the 1890's for improving technology's ability to meet society's needs on a global basis. He felt this would help eliminate the need for warfare.

Although many parts of this "World System" were patented, Mr. Tesla was always secretive about the principles involved. Devices built today according to patent descriptions are seldom as efficient as those made by Tesla himself. Another modern problem is that few experimenters study and apply the details recorded in Tesla's notes.

What follows is a description of Tesla apparatus, tips on construction, and an introduction to the many frontiers which are still open to the electrical experimenter.

Experimenters should take care not to build tuned circuits which broadcast in the AM radio frequency band. An ideal frequency band would be the license-free experimental 90-160 kHz band. Many services including aircraft beacons and ship channels operate from about 200 to 500 kHz. A large, radiating Tesla coil could interfere with these services. The Federal Communications Commission has the power to fine violaters. If in doubt, test your laboratory area by checking for interference on a nearby radio and television while you experiment with your Tesla coil. Working inside a small room with aluminum foil covering the walls and ceiling (creating an electrically grounded Faraday shield) will cut down on some of the electromagnetic waves radiated from the laboratory.

For safety's sake, provide adequate ventilation to remove ozone produced by the apparatus. Work on active electrical parts with one hand in your pocket so as to reduce the danger of accidental shocks. Keep floors dry, ground adequately at the necessary points and wear heavy insulated shoes. Tesla's shoes had one inch thick rubber heels and soles added for protection.

NEVER TOUCH THE PRIMARY OF A WORKING TESLA COIL! Primaries can carry high 60 cycle voltages at high currents. THIS CAN BE LETHAL!

Continuing our description of the Tesla coil transformer, an outer primary coil is wound with a few turns of large gauge wire, while the inner secondary coil is wound with many turns of small gauge wire. This transformer is usually driven by a 60 cycle, high voltage source. Depending on the uses for which the Tesla coil is designed the secondary coil assumes different shapes: the conical shape, large at its base and tapering to a point at its top high voltage terminal and most familiar, the cylindrical coil.

The main innovation Tesla introduced to the design of induction coils in general involved tuning his coils to quarter wavelengths, which greatly simplified construction problems. This means that one end of the secondary remains electrically inactive, while the other end swings up to the maximum potential. This overcame the problem of how to insulate the high voltage secondary coil from the lower voltage primary coil that energized it. This method of tuning enabled the secondary "dead" end to either be connected directly to the primary and/or to the ground, a good safety factor.

One type of Tesla coil circuit is shown in figure one.

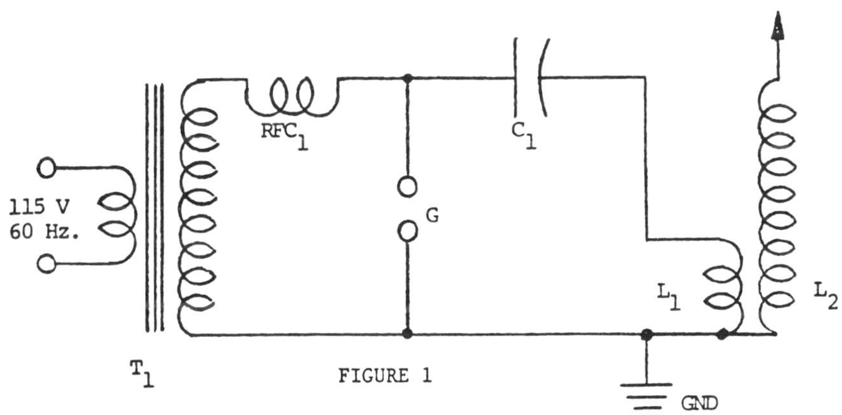

FIGURE 1

A transformer steps up the 115 volt 60 cycle current to several thousand volts which is introduced into a resonant circuit (C1-L1) where it produces a potential dependent on the power input, capacitances, inductance and spark gap setting of the circuit. Power develops in the Tesla coil as follows:

1. The step-up transformer charges the condenser(s) with energy.

2. The condenser (C1) reaches a peak voltage, the spark gap (G) breaks down acting like the closing of a switch which joins C1 & L1 temporarily in parallel. The condenser C1 swaps its energy with L1 back and forth until the energy is consumed. Some energy is used to produce high potential currents in the secondary winding L2 by magnetic and electrostatic coupling.

3. When the current in this resonant circuit stops flowing through the spark gap, the step-up transformer (power supply) begins to recharge the condensers again for another cycle. RFC, is a radio frequency choke which blocks high frequency currents (but is "transparent" to the 60 cycle power), and prevents them from damaging the power supply.

The oscillations produced in resonant circuits containing inductance and capacitance have the frequency:

$$f = \frac{1}{2\pi\sqrt{L_1 C_1}}$$

where f = frequency in cycles per second
L1 = circuit inductance in henries
C1 = circuit capacitance in farads

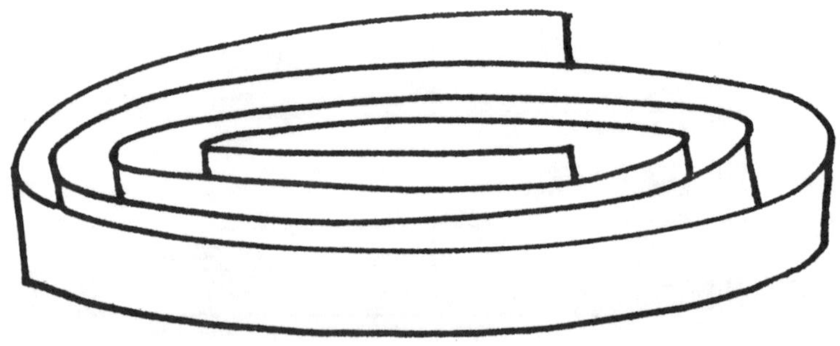

PANCAKE INDUCTOR

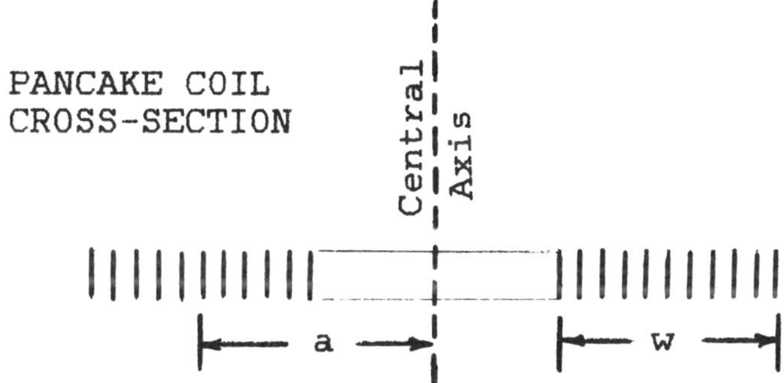

Assume that we want a resonant frequency at 400 kilocycles. Using the formula or tables for the L1-C1 value, conveniently found in HANDBOOK OF CHEMISTRY AND PHYSICS 34th edition, the radio formula section gives the answer. Page 2761 for 400 kilocycles has a value of 0.1583 for the product L1-C1 where L1 is given in microhenries and C1 is given in microfarads. The shape of the primary inductance coil L1, determines the formula for calculating its inductance. For a flat pancake spiral, the inductance L1 in microhenries is approximately given as:

$$L_1 = \frac{a^2 \times n^2}{8a + 11w}$$

where a = average radius in inches as measured from the central axis to the middle of the winding
n = number of turns in coil
w = width of the coil in inches

For example, if a = 8 1/2 inches and n = 10 turns and w = 4 1/2 inches then

$$L_1 = \frac{(8.5)^2 \times (10)^2}{8(8.5) + 11(4.5)}$$

$$= 61.5 \text{ microhenries}$$

$$\text{therefore, } C_1 = \frac{0.1583}{61.5 \text{ microhenries}}$$

$$= .0026 \text{ microfarads}$$

The formula for calculating leyden jar capacitance is given in STANDARD HANDBOOK FOR ELECTRICAL ENGINEERS 6th edition, section 2-139 on page 93.

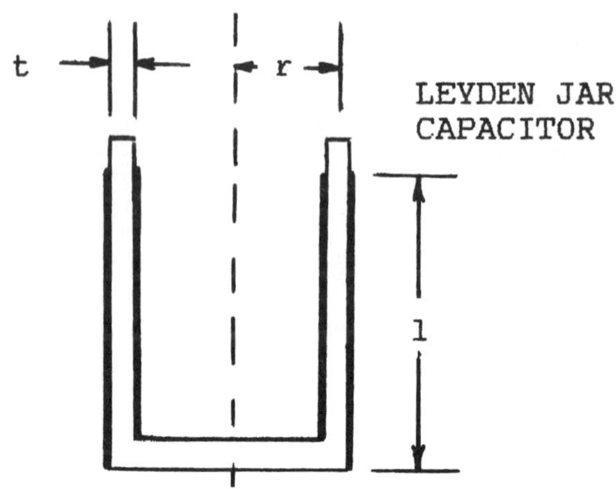

$$C_1 = \frac{.0884\ k(\pi r^2 + 2\pi rl)}{1,000,000\ t}$$

where C_1 = capacitance in microfarads
k = dielectric constant
r = jar radius in centimeters
l = height of the jar portion used in centimeters
t = thickness of the jar wall in centimeters

Be sure the dimensions of the jar are in centimeters. If a 1 gallon glass jar has the dimensions l = 23cm and r = 7.5cm and t = 0.32cm (dielectric constant for glass being approximately 6) then:

$$C_1 = \frac{.0884(6)[\pi(7.5)^2 + 2\pi(7.5)(23)]}{1,000,000 \times (0.32)}$$

$$= .002 \text{ microfarads}$$

TYPICAL DIELECTRIC CONSTANTS

material	k - dielectric constant
plate glass	6.8-8.4
window glass	8.0
Pyrex	4.1-6.1
hard rubber	3.0
mica	6.5-8.0
mineral oil	2.7
bakelite	5.4-5.8
polystyrene	2.5
porcelain	6.5-7.0
shellac film	4.0
spar varnish	4.8-5.5
vulcanized rubber	3.2-3.9
beeswax	2.9
paraffin	2.5
wood	2.0-5.2

Since the capacitance required for the circuit is 0.0026 microfarads, it will take one one-gallon jar measuring approximately 6 inches in diameter by 10 inches tall with a wall thickness of 1/8 inch. If 5 gallon plastic pails or plastic wastepaper baskets are used instead of glass, the dielectric constant will be reduced from 6 down to a value of 2.

Using Tesla's 1/4 wavelength principle, we can calculate the length of the wire needed for the secondary coil. Since the resonant frequency of the primary circuit is 400 kilocycles, the wavelength equals the speed of light divided by this frequency or 750 meters. 750 meters divided by 4 gives 187 meters as the total length of wire needed in the secondary coil. This is about 615 feet. If the secondary coil is in the form of a long cylinder, then for a diameter of ten inches, we would need about 235 turns.

Such calculations are rough approximations because the type of metal used for windings, spacing of windings, length of hookup wires, etc., will influence both capacity and inductance of the entire circuit. These values are for the chosen frequency of 400 kilocycles only.

It is often assumed that the voltage output of the secondary coil is simply determined by the ratio of turns between primary and secondary coils. This is not true. Very high voltages have been produced with secondary coils of relatively few turns.

The general theory of oscillation transformers of high frequency was treated by A. Oberbeck (1895), V. Bjerknes (1895), and P. Drude (1904). The conclusion was that the ability of transformers like the Tesla coil to efficiently generate high voltages depends largely on the RATIO OF THE CAPACITIES in the primary and secondary circuits (of course, inductive and electrostatic coupling between the circuits is also important). Therefore, it is necessary to keep the distributed capacity of the secondary coil, which is determined by coil shape and spacing of windings, as low as possible.

But here is a paradox. The distributed capacity together with the natural inductance of the secondary forms an L-C circuit with its own natural frequency. By tuning the Tesla primary coil to the same frequency as the secondary coil, the generated secondary voltage will be far greater than simply the turns ratio --- the actual voltage produced being dependent on coil "Q" (or quality factor). "Q" can be maximized by reducing coil resistance. The voltage generated can be so high that arcing BETWEEN windings can occur! In other words, intra-winding capacity is needed so that the secondary will resonate, but spacing the windings far apart to reduce arcing (which we want) also reduces the capacity (which we may not want). A tradeoff between these conflicting parameters is needed. When working on new Tesla coil design, several attempts at winding the secondary may be necessary before the optimum winding spacing is found.

An unfortunate tendency in modern coil design is to make the secondary coil of small diameter and long length with windings of very fine wire. This should be avoided because this reduces the efficiency and produces corona leakage problems. Efficiency also is reduced by coupling too closely, that is, by placing primary turns closer to the secondary windings. Overcoupling prevents the circuit from resonating at maximum efficiency.

A few older designs are given here to help offset these design flaws.

The drawings shown in this section date from the approximate years 1914 to 1930. Types of coils known as Oudin or Oudin-Tesla are those hybrid designs which incorporate some of Tesla's designs with those of Oudin, another experimenter working on high frequency currents for medical purposes. Therefore, some of these designs have a secondary coil which runs horizontally (through a centered primary coil) with "live" terminals at each end. These would have to be tuned to ONE-HALF wavelength. Even so, the construction techniques are useful.

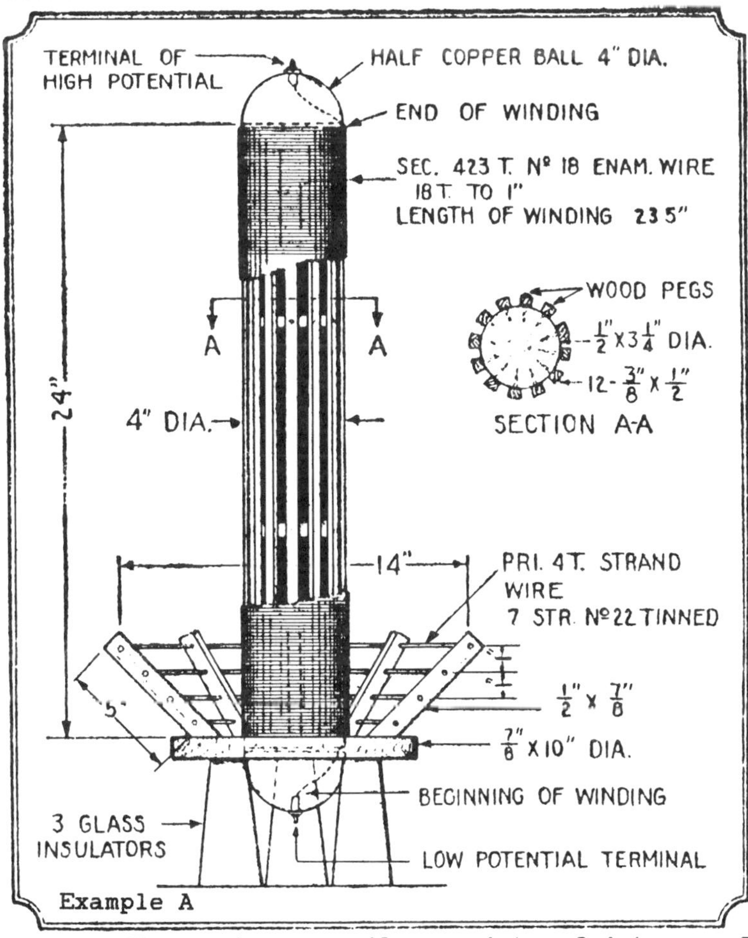

Example A

The primary coil consists of 4 turns of #22 stranded wire. Notice that NO nails are used in constructing the wooden support skeleton. Normally the wooden cylinder for the secondary is turned on a lathe to give a rounded surface for the windings. The secondary consists of 423 turns of #18 enameled wire with 18 turns to the inch, and the total length of winding is 23 1/2 inches long.

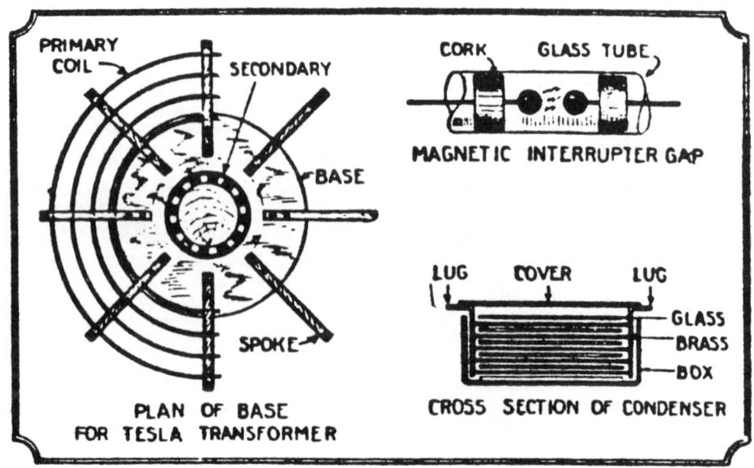

Here is shown a cross-section of the Tesla coil with the plan of the base. A magnetic interrupter gap is shown also and the cross-section of the very substantial condenser required for this apparatus.

Connections to the primary may be in the form of alligator clips which can be clipped at points along coil to facilitate fine tuning. The term "magnetic" interrupter gap means that a magnetic field is used to blow out the arc across the spark gap. This will be discussed in another section.

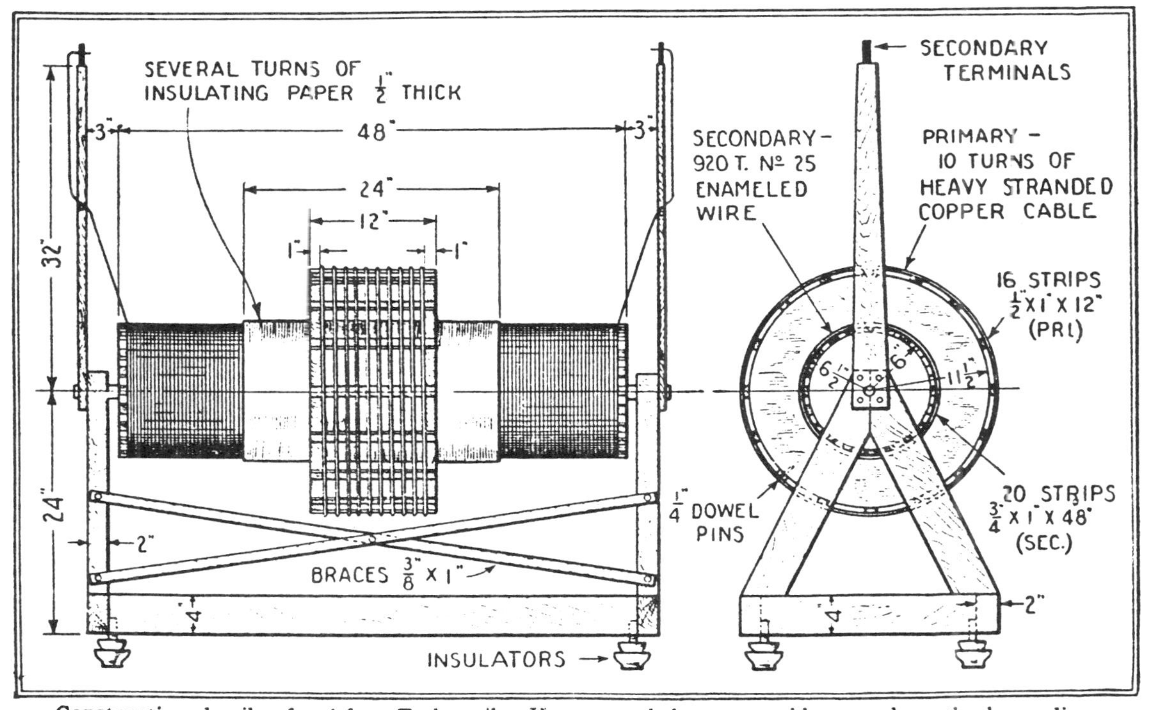

Construction details of a 4-foot Tesia coil. Here stranded copper cable wound on the large diameter drum forms the primary. The secondary is wound on the drum of smaller diameter.

Example B

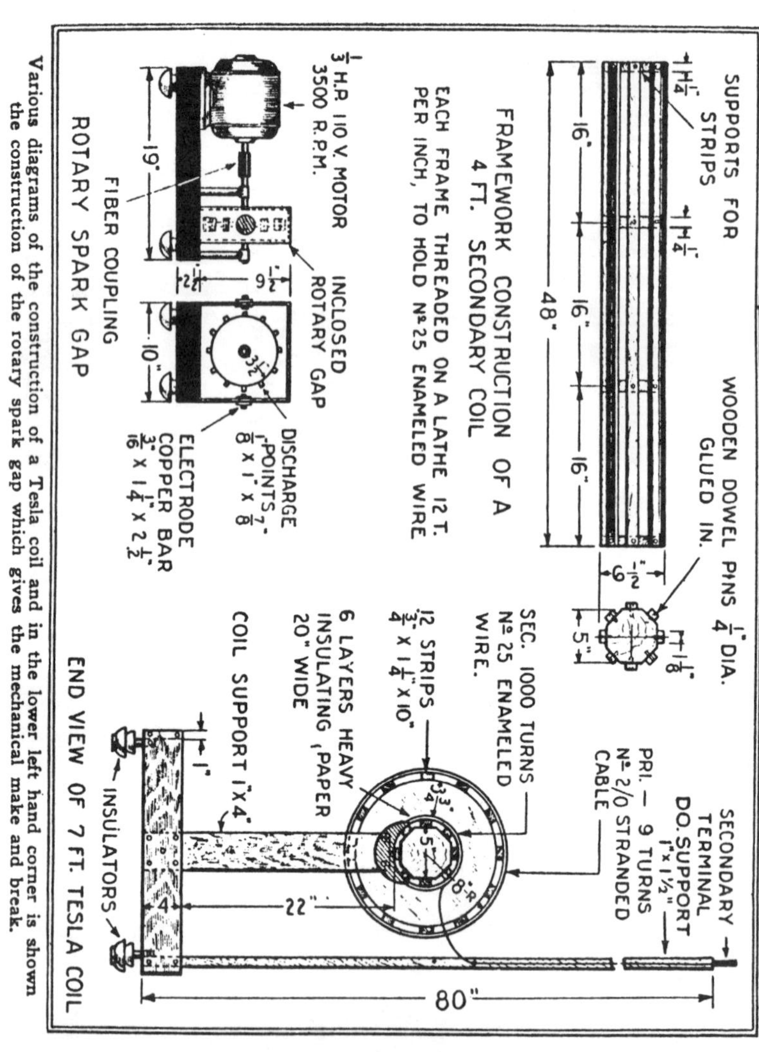

Various diagrams of the construction of a Tesla coil and in the lower left hand corner is shown the construction of the rotary spark gap which gives the mechanical make and break.

Example B is a Tesla transformer which gives a 54" spark. Note that the secondary has two "live" ends. The reader may use, of course, waxed cardboard tubes or PVC pipe for coil forms. Any form should be thin walled and should not absorb moisture; this will improve transformer efficiency. The rotary spark gap greatly improves the flexibility and the efficiency of the Tesla coil and is recommended for all units operating at more than 250 watts.

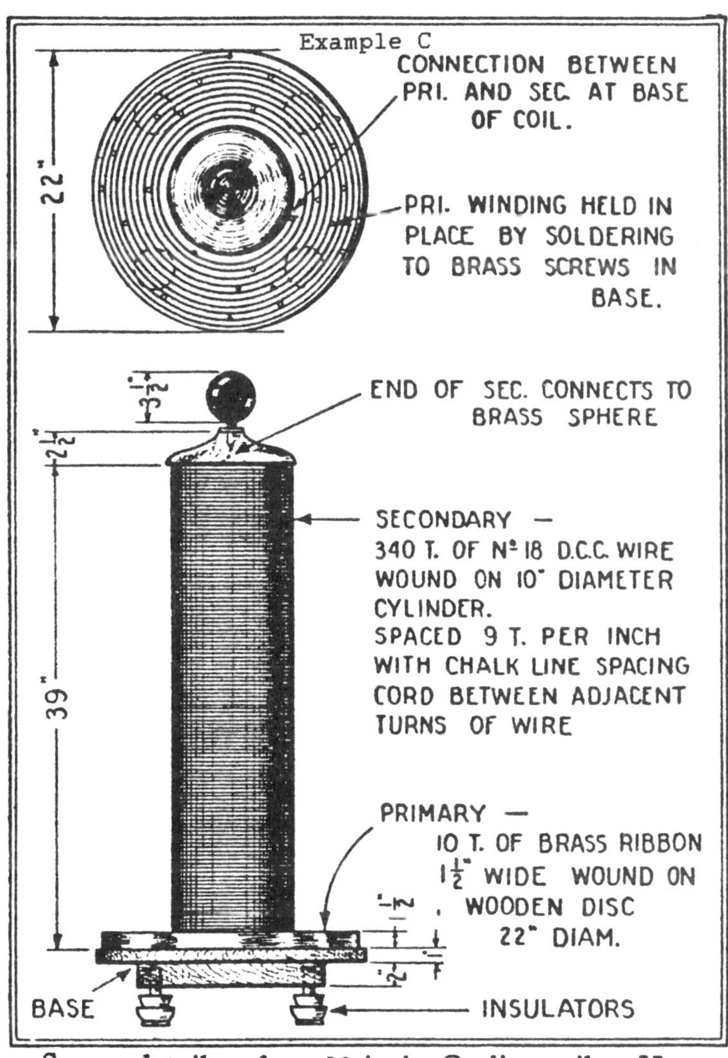

Some details of a 39-inch Oudin coil. Note the primary winding of brass ribbon.

This unit includes the flat pancake ribbon spiral as a primary coil, often used by Tesla. The secondary consists of 340 turns of #18 double cotton covered (DCC) wire which is spaced with chalk line cord to give 9 turns per inch. This type of wire is not available today, so heavy enamel coated wire may be substituted. The metal ribbon needed may be cut from copper sheeting in strips and soldered together using lap joints.

Example D

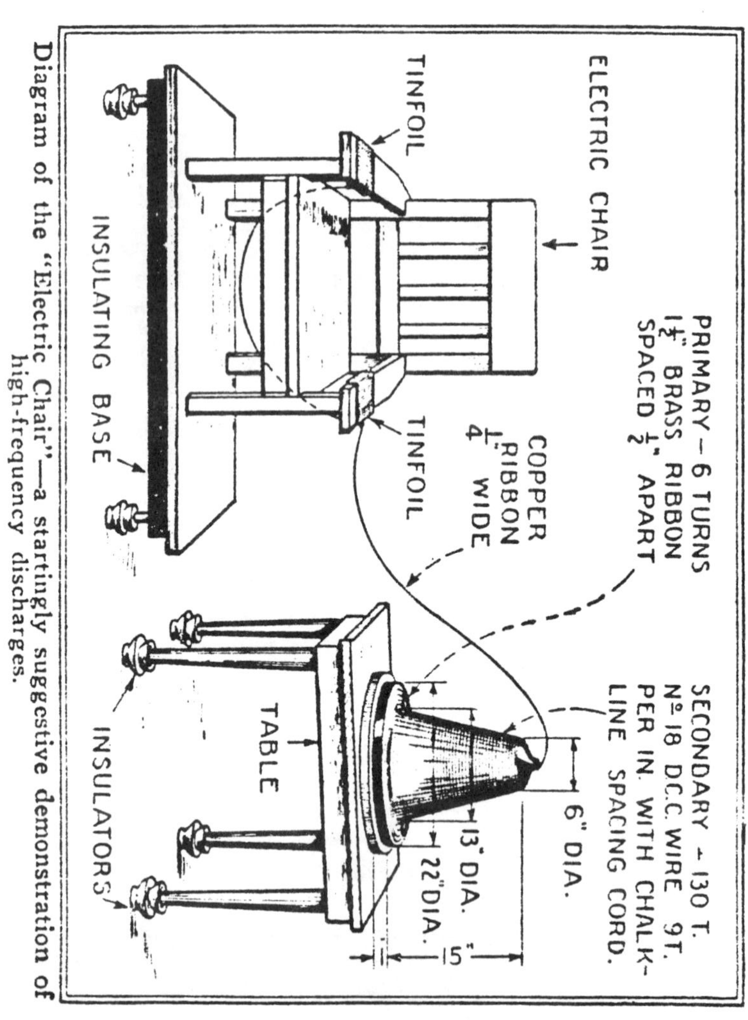

Diagram of the "Electric Chair"—a startingly suggestive demonstration of high-frequency discharges.

Here we see the secondary in the form of a cone because as the height increases above the primary coil so does the voltage; the shape therefore increases the distance between primary and secondary windings and helps reduce corona leakage.

Example E

An Easily Constructed Oudin Coil

By R. C. Hitchcock

WHILE the amateur is obliged to abandon his radio transmitting set, he can easily construct a high frequency oscillator which may be energized by his transmitting outfit.

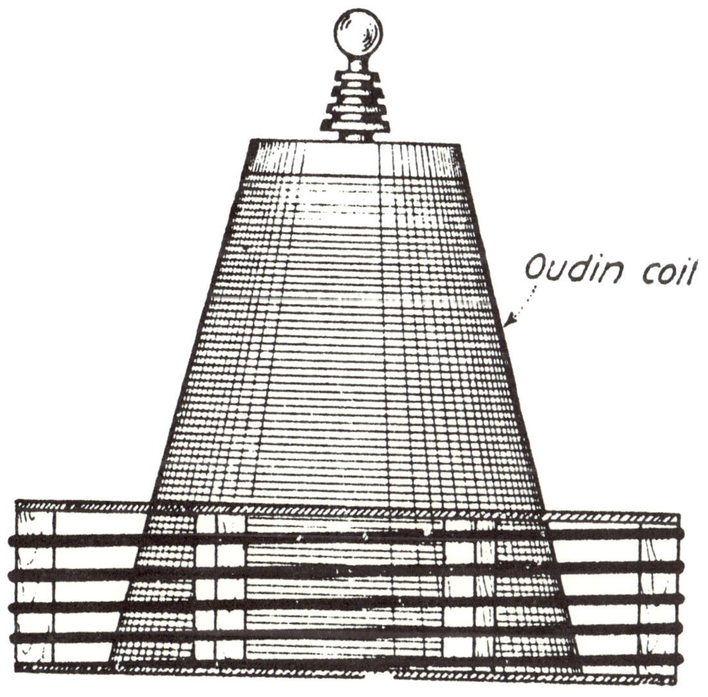

Figure 1—Constructional details of the Oudin coil

Many interesting and startling experiments may be performed.

The coil shown in figure 1 has been successfully operated on a ½ kw. wireless transformer, using a glass condenser and straight gap. The straight gap gave high frequency sparks of greater length than a rotary gap.

The core for the secondary can be made of several pieces of cardboard, cut like the pattern and glued together, making a tube 18 inches long, 9 inches diameter at the small end and

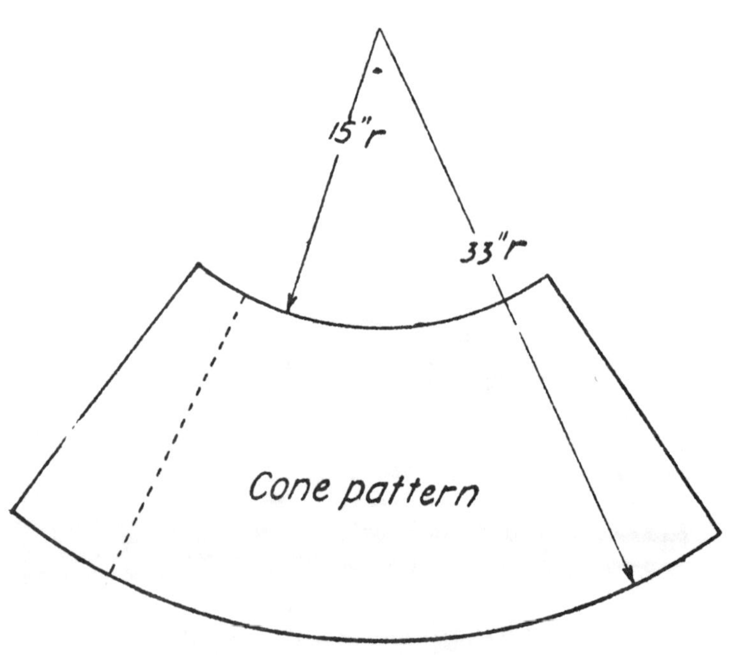

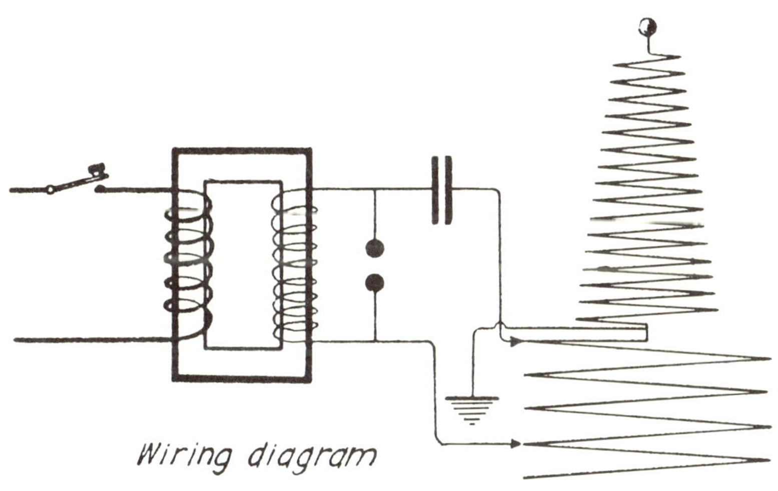

Wiring diagram

18 inches at the other. Two wooden pieces are turned to fit the ends, and the assembly then mounted between lathe centers for winding. No. 26 wire is preferred and the turns should be spaced with a string which may be left on permanently. When the coil has been wound it should be painted with hot paraffine or some good insulating material.

The primary is wound on a six-sided figure about 24 inches outside diameter, depending on the width of the supports. Slots for No. 8 wire are provided, six turns being sufficient for the set under consideration.

An insulator, which may be turned of wood, is mounted on the secondary

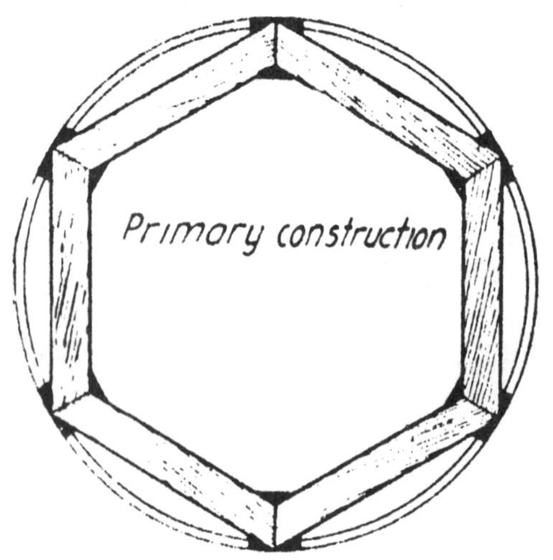

Figure 2—Primary construction

to support a brass ball from which the sparks are drawn.

The secondary and primary are grounded at one end and the other connections made as shown in figure 2.

Sparks drawn into metal objects held in the hand cause no shock whatever.

As the wiring diagram shows, the secondary coil is joined to the primary coil, and is a continuation of it. This is a characteristic of auto transformers of which the Oudin coil is a type. Some Tesla coils use the same connections. The generous diameter of the secondary coil increases its inductance value while the conical shape reduces the distributed capacity of the secondary, which is desirable.

A MINIATURE TESLA COIL.

Most owners of small induction coils have at some time or other wished that a Tesla coil giving results could be built to run on their apparatus. This article describes a Tesla coil made to work with a one-quarter inch spark coil.

Example F

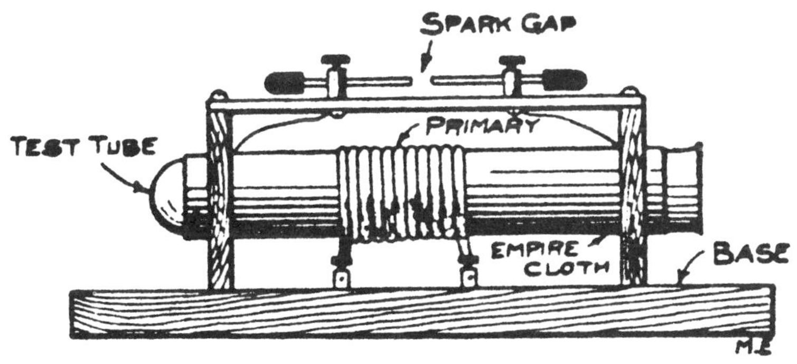

Make a base 8x3x½ inches, and two uprights two inches square and one-quarter inch thick. Now get a test tube 5¾ inches long, inside diameter three-quarters inch. A cardboard tube of the same dimensions will do. Through each of the uprights drill a hole large enough to let the test tube slip through. Starting one-half inch from the end of the tube, wind on about 135 turns of No. 31 single silk copper wire, spacing the turns 1/32 of an inch apart. About one-half inch from the other end of the tube stop winding. Shellac the wire and your secondary is finished. Now put on a strip of Empire cloth 5½ inches wide, winding on three layers. At this stage the uprights may be slipped onto the test tube and secured to the base 4¼ inches apart. The holes in the uprights must, of course, be large enough to let the test tube, with the wire and cloth, through. The

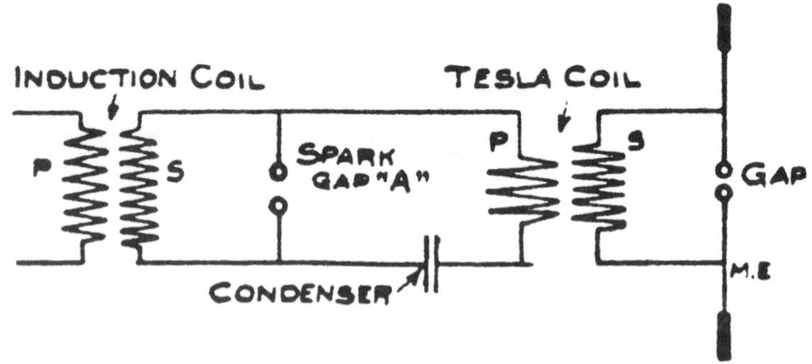

primary, consisting of eleven turns of No. 18 rubber covered copper wire, is now wound over the Empire cloth; 20 turns of No. 20 double cotton covered wire will do as well. Nail a strip of wood on top of the uprights, and by means of two binding posts fastened to it, make a spark gap. Connect the secondary wires to these posts; the primary wires terminate in binding posts fastened to the base.

The above coil, run in conjunction with a one-quarter inch spark coil operated by three dry cells, and a single small Leyden jar, gave a spark nearly one-half of an inch long. The brush discharge at the spark rods in the dark, is about one and one-half inches long when no spark is passing. The spark varies in color and thickness with the materials used in the spark gap. When allowed to jump to the hands it produces no sensation whatever. Use pointed zinc or brass rods in spark gap A in the diagram.

Contributed by

CARL DREHER.

This miniature Oudin-Tesla coil shows how a small and inexpensive apparatus can be built from ordinary supplies. Heavily enameled copper wire can be substituted for the cotton covered wire. The upright supports for the test tube should be fastened with dowels not nails. The pyrex test tubes also make fine leyden jar condenser banks.

Apparatus For Vaudeville Act (1914)

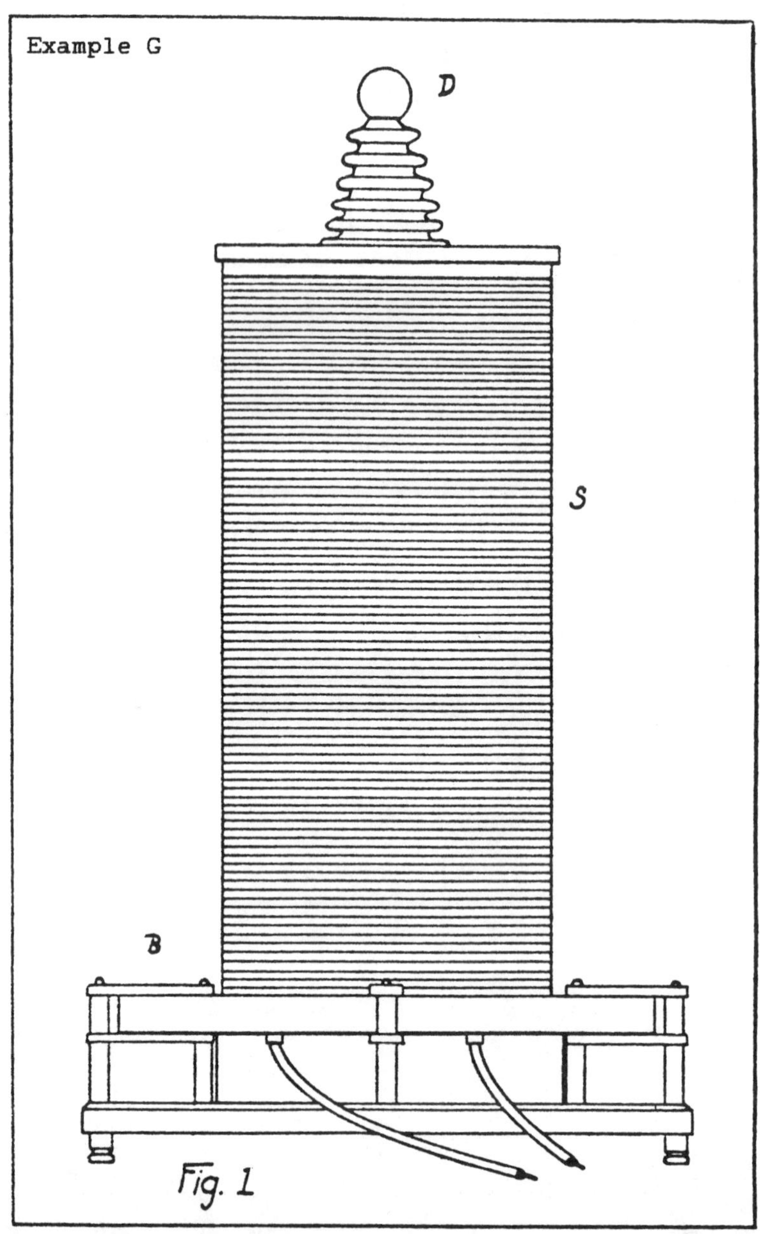

Special Transformer for Producing Current for Electrical Stage Act.

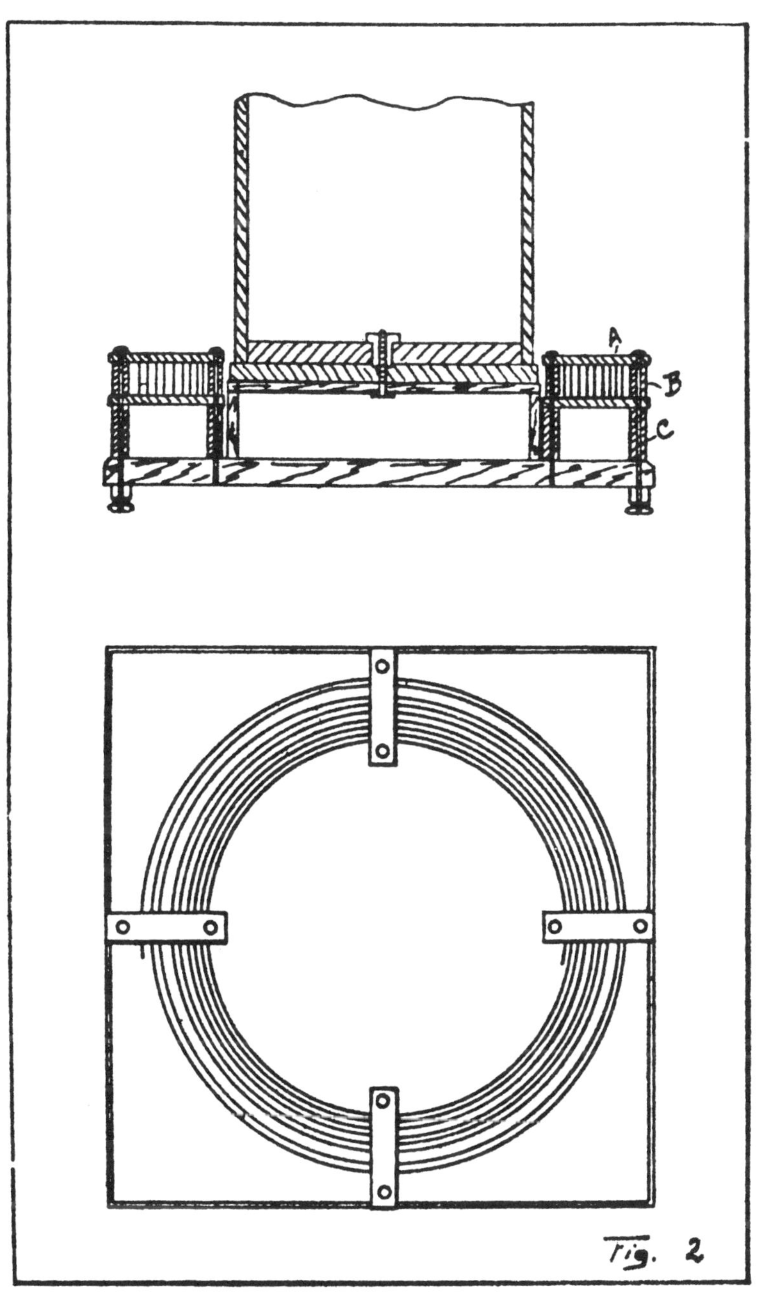

Plan and Section of Primary.

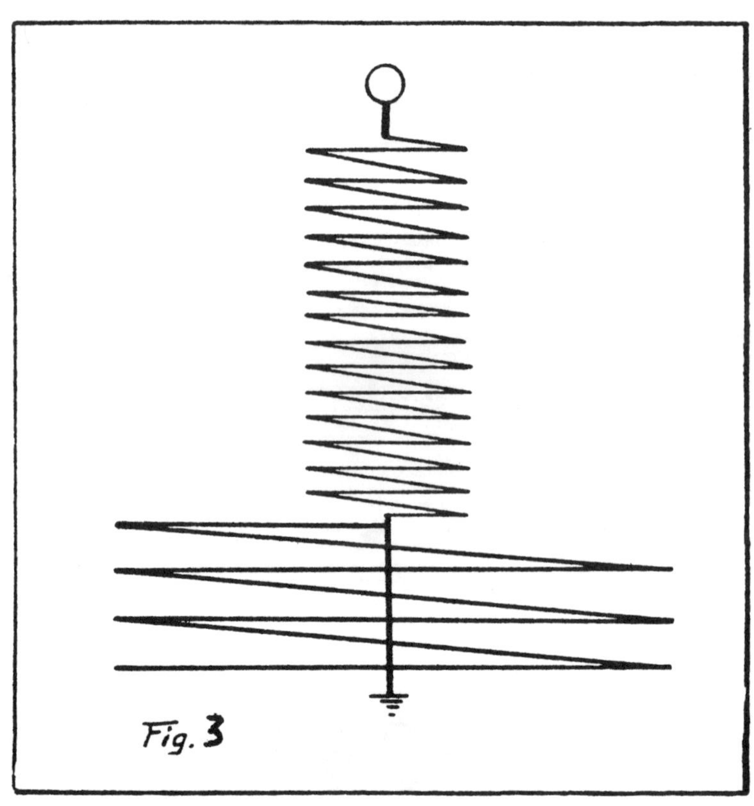

Connections for Complete Transformer.

The primary (B) in Figure 1 is a flat pancake consisting of 9 turns of heavy copper ribbon 2" wide and wound in a spiral. A double thickness of 3/16" thick rubber belting is also wound in to space the turns apart. This coil (Fig. 2) is elevated a few inches so as to permit the two wires shown at the bottom to be slid along the spiral turns for tuning the primary. The secondary coil(s) (Fig. 1) is composed of 600 turns of #22 double cotton covered copper wire, 12 turns per inch spacing, which is wound on a cylinder 20" diameter by 50" tall. The top terminal holder is turned from wood, and a bolt passes through this to join the secondary wire with the hollow brass ball on top. Figure 3 shows the secondary joined to the inside primary turn and this point is grounded. Properly tuned, this coil gives a 5 foot long spark.

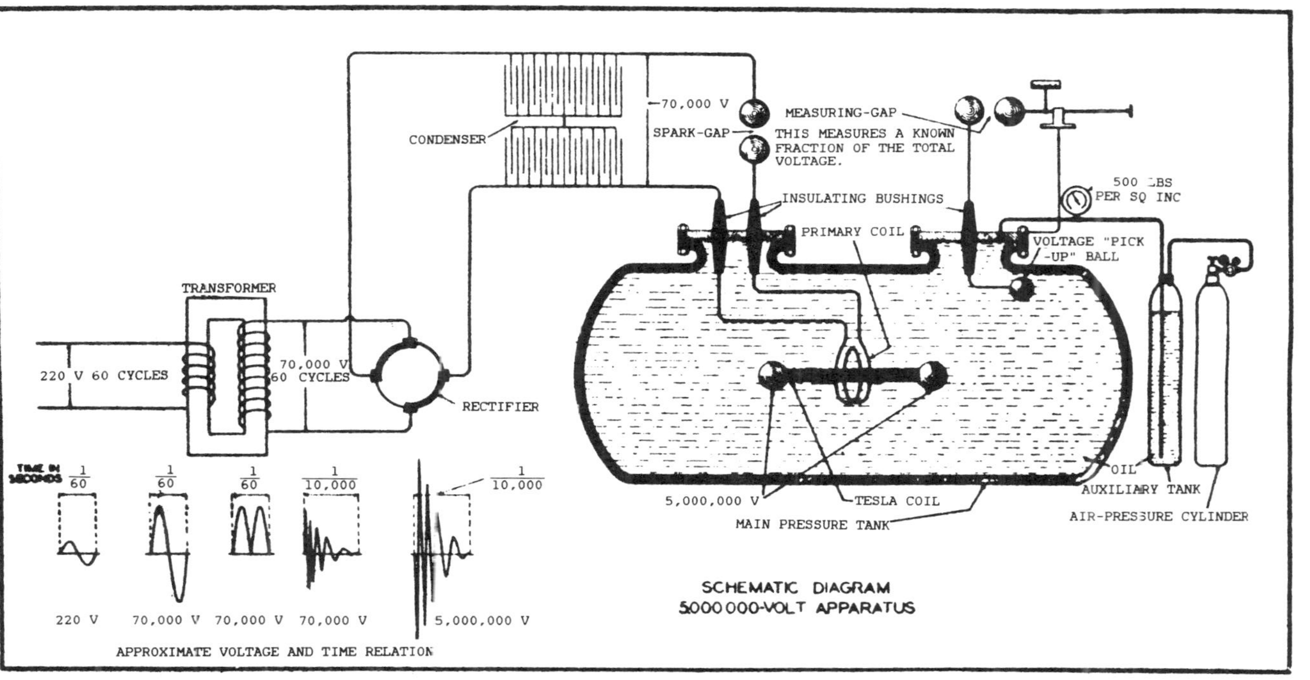

Fig. 1. Schematic diagram Tesla coil in oil under pressure.

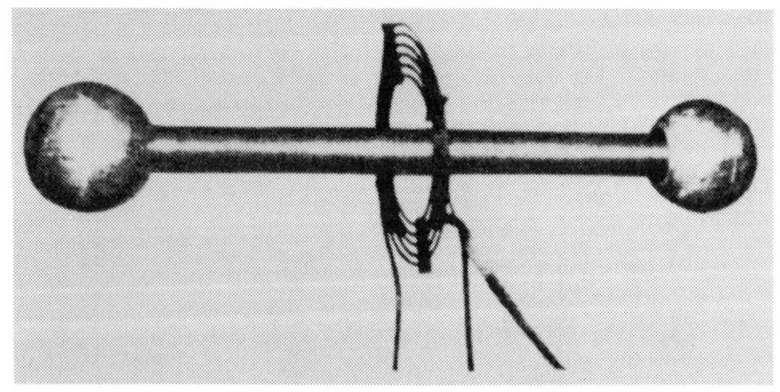

To reduce corona leakage the entire apparatus is submerged in light transformer oil which may be pressurized. The condenser consisted of sixteen glass-plate-in-oil types, each with a capacity of 0.038 microfarads and each rated at 15,000 volts. The primary coil is a flat pancake of 6 turns of copper tubing. The secondary is wound on a pyrex glass tube 8 cm. diameter by 1 meter long with spherical end caps of spun-zinc 25 cm. in diameter. Wire size was #40 B.S. gauge heavily enameled wire 5000 to 7000 turns without spacing. The trade name of the oil used was "Transil 10C" which had a dielectric strength of 46,000 volts per millimeter thickness.

Example I

Electrical Wizardry at Home

Some Weird Effects That You Can Obtain with a Tesla Coil and Some of Your Apparatus

To begin with, if you have an oscillation transformer the secondary will make a suitable primary for a Tesla or Oudin coil.

To construct a Tesla coil that will give a 10- to 12-in. spark, procure a tube 4 in. in diameter and 16 in. long. This may be made of two tubes 8 in. long and joined together as shown in Fig. 2. Some cereals come in tubular containers which will answer very well for this purpose. The pieces *A* and *B* are of wood; shellac is used to join the parts together. Avoid the use of nails in this construction.

After applying two coats of shellac to the secondary tube, wind it closely within ½ in. of either end with

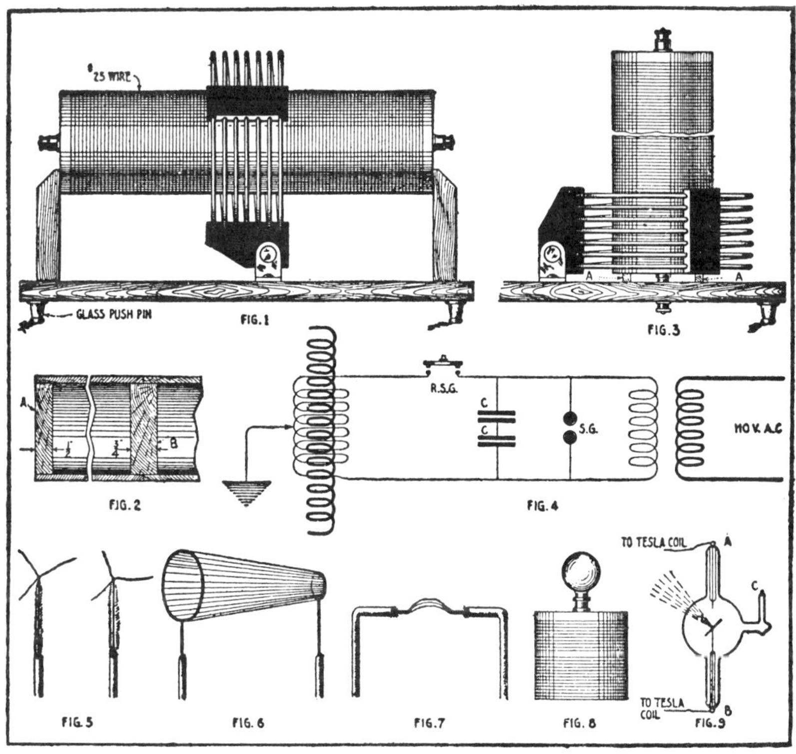

Details in the construction of a Tesla or Oudin coil and diagram of the hook-up for the experiments shown in the lower part of the illustration. These experiments as well as X-ray photographs may be made by the use of this coil

double cotton-covered No. 25 wire. Then give two coats of shellac to the windings, being sure the first coat is absolutely dry before applying the second.

The base and uprights should be made of well-seasoned wood which should also be shellacked. Make the dimensions to suit the individual requirements.

Fasten four glass push-pins to the base as shown in Fig. 1. These serve as insulators.

In Fig. 3 is shown how the above-described Tesla coil with a few alterations can be converted into an Oudin coil. The three small blocks *A* are used to support the secondary.

The hook-up for a Tesla coil is shown in Fig. 4, and the abbreviations used are as follows: R. S. G., rotary spark gap; C and C, condensers; S. G., safety gap.

Various experiments are shown in Figs. 5, 6 and 7. The two wires in Fig. 5 are connected with the binding-posts of the secondary and are left to project vertically in the air. Long streamers wave about, producing a weird effect.

The experiment in Fig. 6 produces a cone of light. Two wire hoops, one 12 in. in diameter and the other 3 in. in diameter, are connected with the secondary, and the lead wires are so bent that the hoops are separated 5 or 6 in.

When a gap is made as shown in Fig. 7, a spark 10 to 12 in. is obtained. Of course all these experiments should be made in the dark.

If two metal disks about 1 in. in diameter are provided and one attached to each of the connection leads, a brilliant flow of light will be produced from their edges. For another experiment, suspend two metal rods from the ceiling or other support so that they will hang about 2 in. apart. Connect these to the leads. Sparks will start at the bottom and run to the top, making a ladder of light.

When the coil is converted into an Oudin coil the bottom binding-post is connected with the lower turn of the primary; otherwise the connections are the same as for the Tesla coil, as shown in Fig. 4. A brass ball 2 in. in diameter, Fig. 8, should be screwed on the top binding-post of the secondary in place of the thumbnut. One of these may be obtained from an iron bed.

An interesting field for high frequency experimenting is in connection with the X-rays. The bulb, Fig. 9, is connected with the secondary leads of the Tesla coil at *A* and *B* on the bulb. The wire *C* is the vacuum regulator, and is operated by bending it over near *A* so a spark will jump while the tube is running.

Good X-ray photographs of the hand or other objects may be made with this apparatus. To take an X-ray photograph, load a plate holder with one plate only, then expose it, holding your hand or other object against the side of the plate holder nearest the plate. The X-rays penetrate the light-proof slide with ease. The time of exposure can be determined only by trial. A good printing negative can be made by holding the plate holder 5 in. from a 6-in. tube and exposing it for 2 minutes.

from HOW TO MAKE THINGS ELECTRICAL

by UPC Book Co Inc 1922

Some samples of Tesla coil hookups are now given:

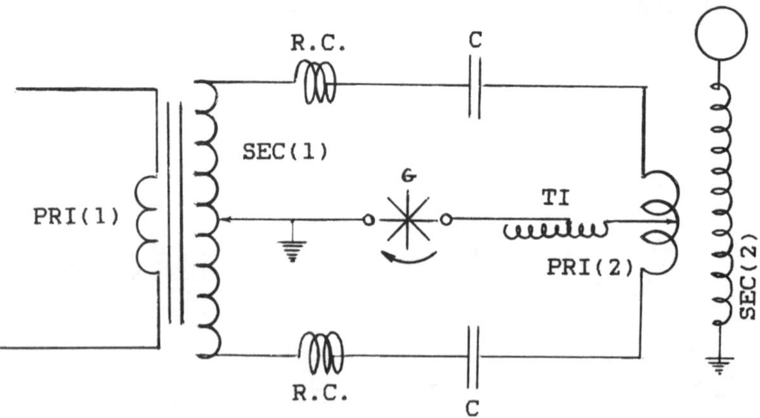

To the left, "PRI/SEC", is the 60 cycle step-up transformer with output approximately 10,000 to 15,000 volts. The secondary is center-tapped to ground. "R.C." is the radio frequency air core choke (a safety measure which must be used), and "C" represents the leyden jar condenser banks. "T.I." is a tuning inductance, multiple tapped (air core) for varying the inductance of the primary coil, "PRI(2)" at the right. "SEC(2)" is the high voltage coil output. "G" is the rotary-gap motor-driven interrupter.

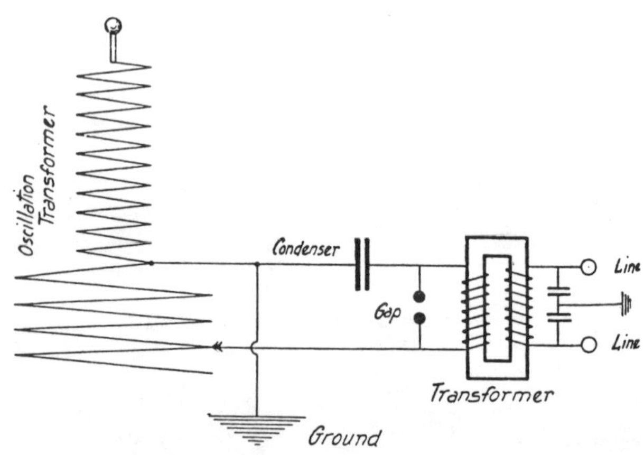

In this circuit, the Tesla coil secondary at the left is a continuation of the primary. This is also known as an auto-transformer. Two capacitors and the central ground are a safety measure for protecting the step-up power supply windings.

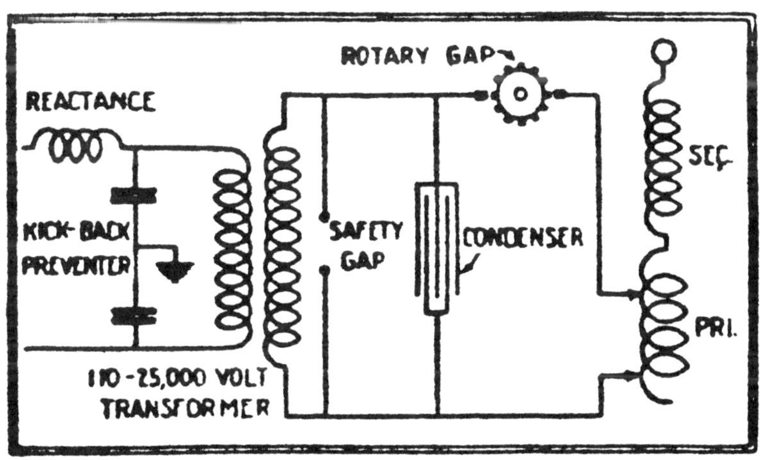

Diagram of the connections for high frequency apparatus as applied to an Oudin coil. Notice particularly the rotary spark gap, giving a mechanical make and break, and also the kick-back preventer grounded to the transformer core.

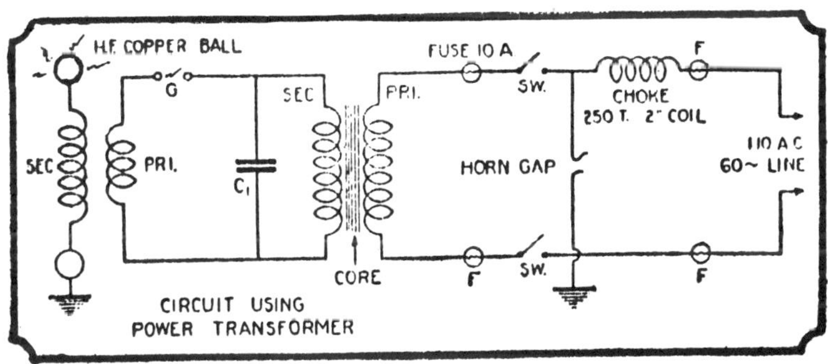

Again, the kickback preventer, reactance or choke and the horn gap are measures for protecting both the step-up transformer and the experimenter from heavy currents.

KICKBACK PREVENTER

The Underwriters' rules call for a protective device in all cases where a step-up transformer is attached to the lighting circuit. An easily made kickback preventer is described herewith. It consists of a fused switch block, choke coil, horn gap and separate fuse block.

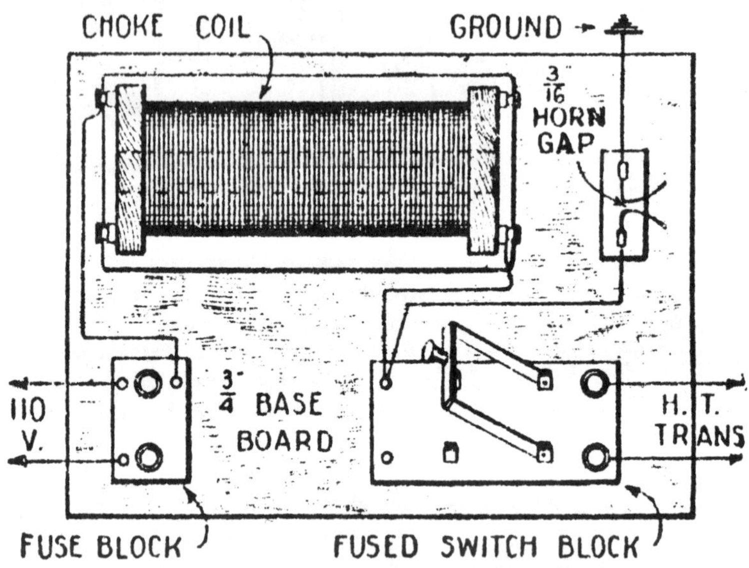

If a step-up transformer is attached to an A.C. lighting circuit, the insurance authorities sometimes require a kick-back preventer. Above is shown one embodying a horn gap on the ground line, and a choke coil. A fuse is required on the switch block.

The horn gap is made from a porcelain base taken off a knife switch. Upon it are mounted two pieces of 3/16" round brass rod, bent horn shape. The air gap should be 3/16". A seven inch piece of 2" diameter tubing may be used for the choke coil. The ends are turned up from pine wood. Four layers of 100 turns each, of number 16 or 18 gauge s.c.c. wire are wound on this spool, each layer terminating at a binding post. This choke coil not only chokes the high tension current which may kick back into the power line but also serves to limit the input to the transformer, which may be regulated by varying the number of layers in the circuit.

In this sketch, "H.T. Trans" (high tension tranformer) refers to the leads going to the primary side of the 60 cycle step-up transformer power supply. Where two capacitors are used for a kickback preventer, each is a high voltage unit having a value of about 0.02 microfarads.

As for the step-up transformer, a general rule for sizing (air core Tesla coils only) is 400 to 500 watts input for each 1 foot of spark length desired. This means 110 volts 60 cycle at 3 1/2 to 4 1/2 amps in drawing capacity. The Tesla coil primary can therefore have very high 60 cycle voltage and dangerous currents. Again, USE CAUTION around the primary circuit. Tesla coils are occasionally powered by pulsed high voltage direct current in which the output of the power supply passes through a high voltage rectifier and filter.

In order to study electrical resonance phenomena, especially with wireless power transmission experiments, Tesla endeavored to produce pure sine waves. This is quite difficult normally since losses through the coils, hookups, condensers, spark gap and corona leakage are high and the C1-L1 circuit can oscillate about five pulses per second, that is about 5 bursts of high frequency waves separated by periods of no oscillation as shown in the oscillograph.

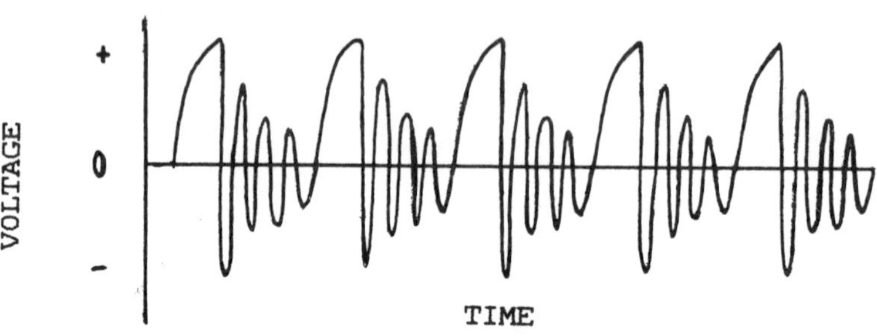

This need for sine waves without distortion, harmonics, etc., was partly solved by principles discovered by Tesla but all the techniques did not show up in his notes. Some of these methods involve direction of coil winding, shape and form of coil windings and rapidly quenched spark gaps or mechanical interruption of the circuit. Each coil form and its direction of winding produced its own unique kind of wave shape. A few possibilities are produced in this publication, and this area is itself a new frontier for experimenters.

One coil shape which Tesla used in his wireless power experiments was the flat pancake spiral. This type of winding tended to have a reduced distributed capacity and was preferred to the cylindrical coil because of the better wave form produced.

The flat spiral was used quite often in transformer experiments in the 1800's. The scientist Matteucci in 1840 published his work with flat spirals.

Inductive action of the Leyden discharge.—Figure 770 represents an apparatus devised by Matteucci, which is very well adapted for showing the development of induced currents produced either by the discharge of a Leyden jar or by the passage of a voltaic current.

It consists of two glass plates about 12 inches in diameter, fixed vertically on the two supports A and B. These supports are on movable feet, and can either be approached or removed at will. On the anterior face of the plate A are coiled about 30 yards of copper wire, C, a millimetre in diameter. The two ends of this wire pass through the plate, one in the centre, the other near the edge, terminating in two binding screws, like those represented in *m* and *n*, on the plate B. To these binding screws are attached two copper wires, *c* and *d*, through which the inducing current is passed.

On the face of the plate B, which is towards A, is enrolled a spiral of finer copper wire than the wire C. Its extremities terminate in the binding screws *m* and *n*, on which are fixed two wires, *h* and *i*, intended to transmit the induced current. The two wires on the plates are not only covered with silk, but each circuit is insulated from the next one by a thick layer of shellac varnish.

In order to show the production of the induced current by the discharge of a Leyden jar, one end of the wire C is connected with the outer coating, and the other end with the knob of the Leyden jar, as shown in the figure. When the spark passes, the electricity traversing the wire C acts by induction on the neutral fluid of the wire on the plate B, and produces an instantaneous current in this wire. A person holding two copper handles connected with the wires *i* and *h* receives a shock, the intensity of which is greater in

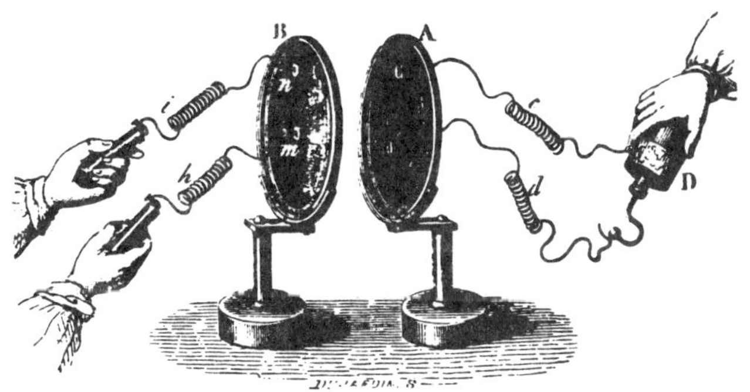

Fig. 770.

proportion as the plates A and B are nearer. This experiment proves that frictional electricity can give rise to induced currents as well as voltaic electricity.

This is what people did for entertainment in the good old days! The tendency is for the spirals to also act as plates of a condenser so that transformation of energy takes place by electrostatic and inductive coupling. Later in 1903, the most interesting experiments performed by Dr. Guilleminot indicated more peculiarities of flat coil spirals:

BY EMILE GUARINI.

Dr. Guilleminot recently published the results which he had succeeded in obtaining with an arrangement for the purpose of analyzing the effects of high frequency obtained with ordinary resonators. He prepared a coil or spiral of copper wire of constant pitch. Through the outermost convolution (Fig. 1) the oscillatory current of two Leyden jars was passed. At the center of the spiral, currents of exceedingly feeble intensity were received.

By modifying the pitch of the spiral and rendering

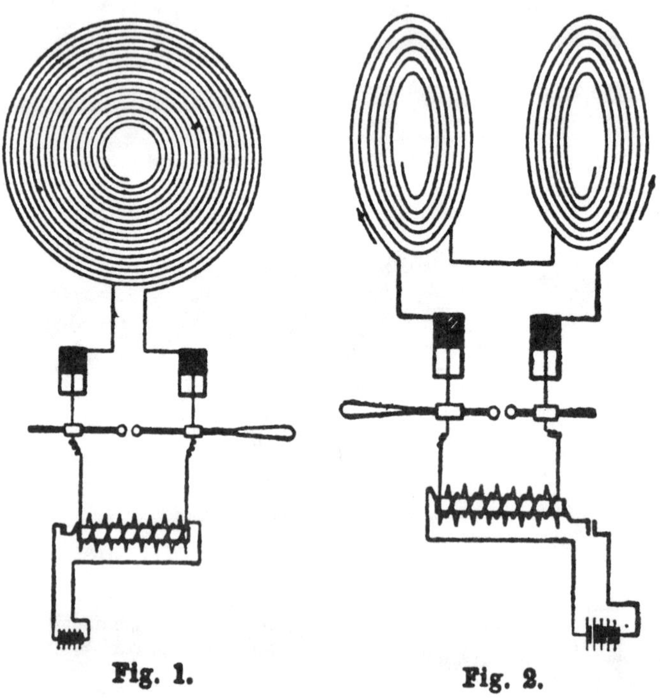

Fig. 1. Fig. 2.

it progressively increasing from the center to the periphery, the effects obtained were more marked. The increase of pitch was based on the difference of the length of the spark which passes between the two adjacent points. This increase was 0.003 mm. between convolutions, an induction coil having a spark of

0.35 mm. being employed at 6 amperes to charge the Leyden jars.

Dr. Guilleminot studied the effects obtained with two spirals. The results were remarkable. An enormous field of action was given to the neighboring resonator. Experiments conducted in conjunction with MM. Radiguet and Massiot proved that if a passive spiral be submitted to the influence of an active spiral, entirely different effects, dependent upon the direction in which the convolutions run, are obtained; if two spirals be connected in multiple (Fig. 2) or in tension, the effects are again entirely different, dependent on the direction of the oscillating discharge in the first spiral. From these experiments it follows that in two spirals the same charging effect can be obtained, either by influence due to winding in the same direction, or by proper connection due to symmetrical mounting; and that it is possible to

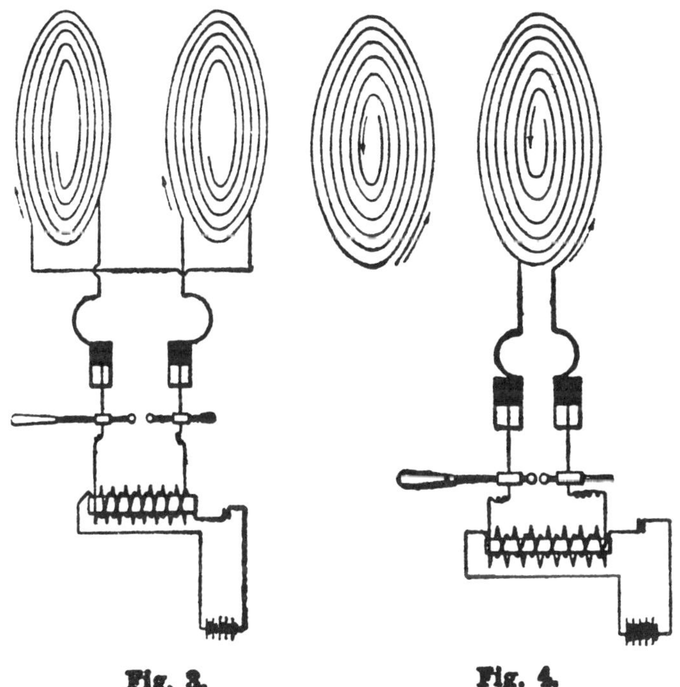

Fig. 3. Fig. 4.

obtain a counter-discharge effect either by influence or by proper connection.

Dr. Guilleminot has also ascertained the effect of combined inverse connection and inverse winding. The results obtained are striking. Superb inter-polar effects were secured. A body interposed between the two spirals glowed at two sides.

By homologous connection and winding feeble current effects were obtained; a body interposed between these spirals emitted flaming streams.

It will be observed that Dr. Guilleminot's spirals enabled him to obtain two entirely different effects, one monopolar and the other bipolar. The therapeutical value of these coils should not be underestimated. Electric shower baths can be taken, which will doubtless have no slight beneficial effect upon the nerves.

In 1904, Dr. Clarence Wright described his experiments on the inductive properties of different coil shapes in the JOURNAL OF THE RONTGEN SOCIETY. His work compared flat spirals, cylindrical, conical, spherical plano-convex and biconvex shapes, each with its own unique inductive values.

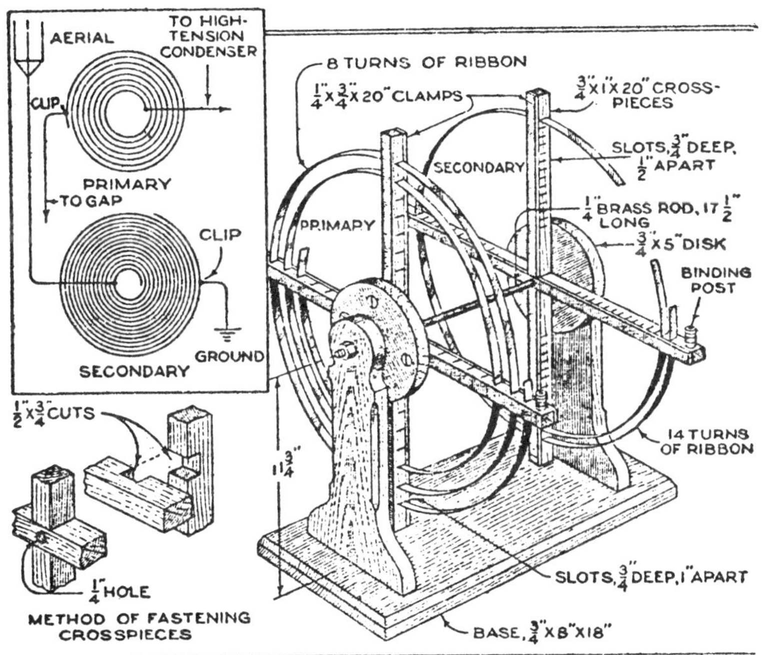

A Simple and Easily Made Oscillation Transformer, of the So-Called "Pancake" Type: All Material, with the Exception of the Ribbon, can be Picked Up around the Radio Laboratory

This picture shows how flat pancake spirals may be constructed. This particular unit is for connection to radio sets, not to a Tesla coil. The ribbon in this case was 1/16" thick and 3/4" wide copper or brass. The use of steel ribbon will greatly affect the inductance of the coil, but it is an option.

In all coil constructions, kinks and sharp bends in windings must be avoided. Where splicing of wire is needed the proper joint is as follows:

The ends are bared by polishing with sand paper, lapped 1/8" and offset to give a smooth transition, then soldered. Never twist wire as a sharp bump results and strains the wires. Cover the joint with a folded paper tab and wind the tab in with succeeding turns as shown.:

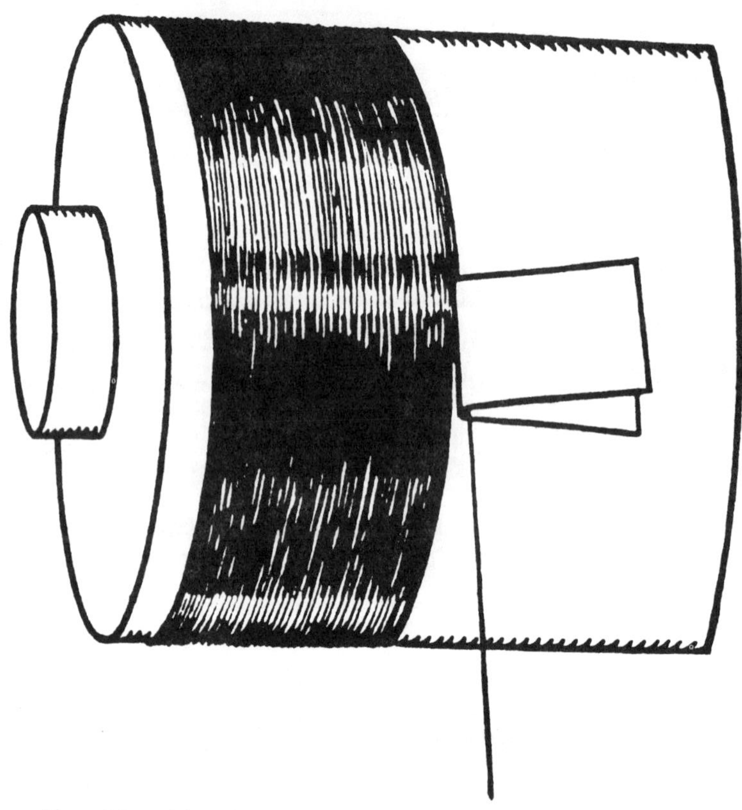

FIG. 98.—Sketch of paper tab folded in to cover joint in secondary wire.

The high voltage terminal on a Tesla coils is often shown too small in size, often just a knob. This can seriously reduce spark length. Generally, when using the hollow sphere shape, it should be at least equal to the diameter of the secondary winding.

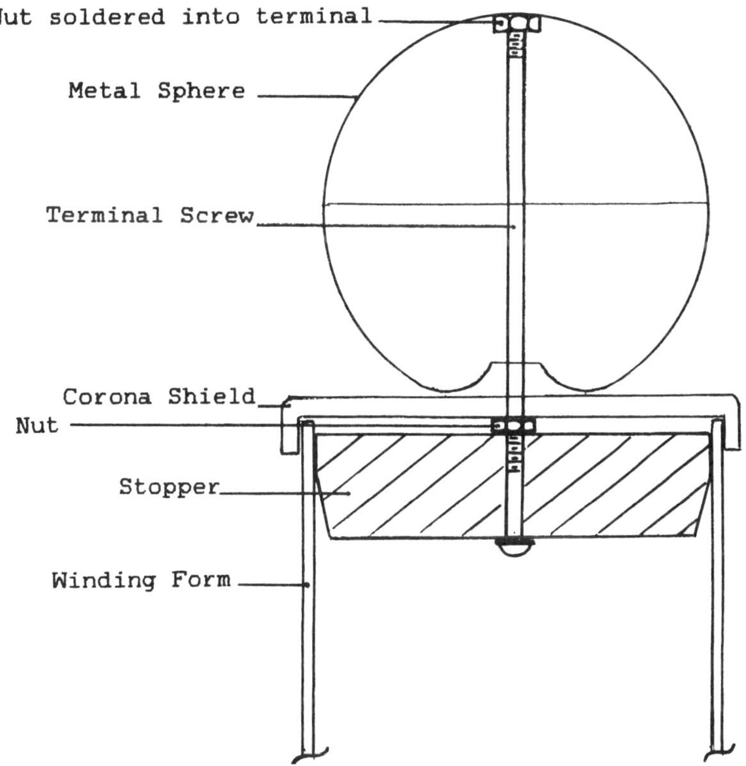

The sketch shows a split sphere inside which at the top is soldered a nut. At the terminal base is the entrance hole, enlarged and turned in to reduce corona leakage. The stopper can be compressed with a nut to fit snugly into the winding form. The corona shield, made of plastic or rubber turns down over the form top, and the secondary wire is fed in through the form to the terminal screw. For large Tesla coils, having spark discharges of 5 to 10 feet, a large manufactured terminal several feet in diameter can be expensive. One simple inexpensive solution which has been proven effective involves making an internal wooden support skeleton of the desired shape, the outer edges of which are notched so as to permit spiraling soft copper tubing over the entire surface by nesting the tubing in the notches. Spacing between turns equal to about twice the tubing diameter is acceptable. At the top of the sphere the tubing terminates by pointing downward into the terminal and becomes the electrical connection and support.

Tesla did recognize the importance of terminal shape and size. He intended to use a large toroidal (doughnut shaped) ring in his wireless power transmission experiments but could not afford to build it. The experimenter should not limit the shape to spheres or oblates but should consider the ring shape, which can dramatically affect spark length.

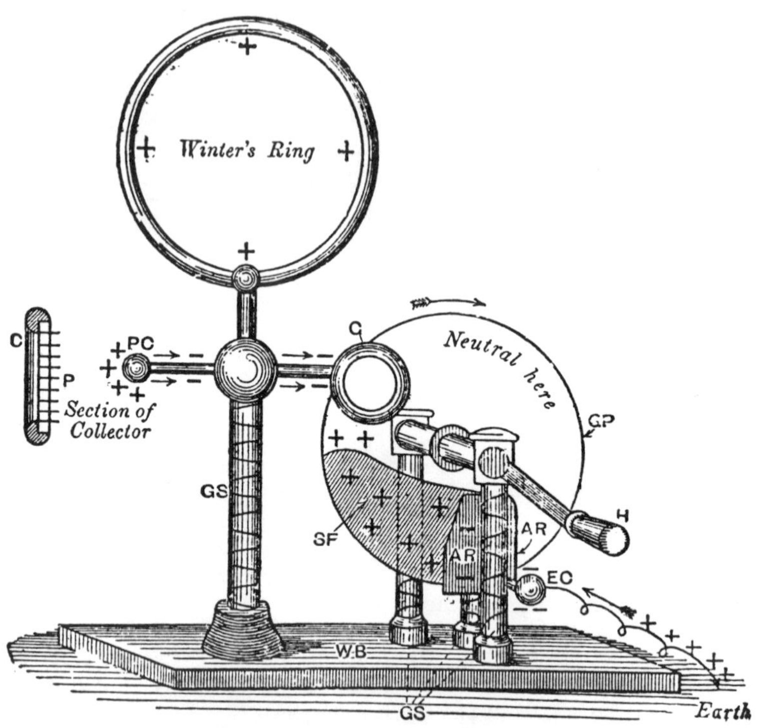

WINTER'S PLATE-GLASS MACHINE.

Historically, in 1856 K. Winter of Vienna displayed a frictional electric machine with a glass plate 3 1/2 feet in diameter having a high voltage ring terminal with a thickness of 1 1/2" and a diameter of 27 inches. Sparks 2 feet long could be drawn from "Winter's ring" which was quite an achievement at that time. But most unusual was that Winter's ring consisted of wood sections, well polished, within which was sandwiched a small wire (27 inches diameter). Thus was formed a terminal having semi-conducting

HEMISPHERICAL TERMINAL
(Copper Tubing)

Support Structure

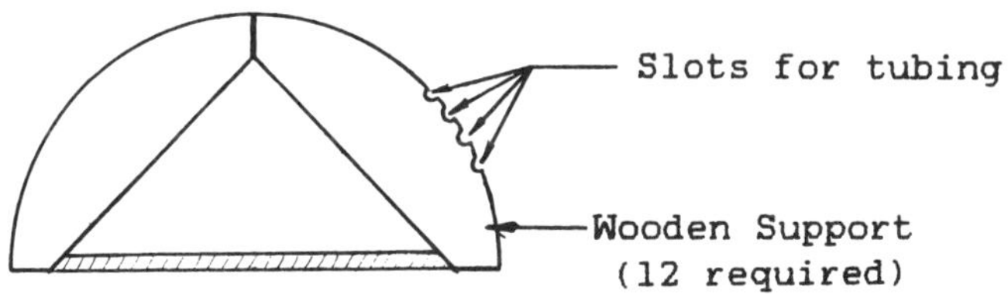

properties! This avenue is completely unexplored. Wood-covered rings would be a fire hazard for the larger units but other semiconductors can be used, cast stone powders having hygroscopic properties being one possibility.

Since Tesla coils, like radios, must be fine tuned to achieve maximum output, each type of output terminal will have its own effect on the capacitance of the secondary coil. Retuning will probably be required. Again, maximum output depends on the ratio of capacities of the primary and secondary circuit (as well as on the effective coupling between primary and secondary).

Leyden Jar Condensers, Sparkgaps & Mechanical Contact Breakers

One of the shortcomings of modern Tesla coil designs has been a lack of adequate means for fine tuning in order to produce the efficient 1/4 wave length oscillation in the secondary winding. This is done to reduce space, weight and simplify construction details. The writer recommends sticking with the more traditional designs developed by Nikola Tesla where replication of his experiments is the goal.

The wet leyden jar, the original electrical condenser, is recommended because it is one means of fine tuning and its capacity is changed by raising or lowering the level of the water within.

The jar of the dry leyden jar shown is pyrex glass 38 cm. high, 11 1/2 cm. diameter with foil coatings inside and out 20 cm. high. Glass thickness is about 4 mm. and the capacity is 0.002 microfarads. Jars of this design can be used up to 100,000 volts D.C. which is more than enough for the usual voltages found in Tesla primary circuits.

Tesla's wet condensers were made using banks of mineral water jugs (lead glass) which were placed in a common tin tank. In place of metal foil, water with a concentrated solution of rock salt formed the "plates".

I recommend the use of Pyrex glass for better performance since Pyrex has higher dielectric strength and does not absorb films of moisture, which is a source of serious leakage.

The cap supporting the charging rod may be of plastic or may be a rubber stopper. Cleanliness of the glass jars is a must and includes even the grease from finger prints. Finally, leakage is reduced by pouring a layer of paraffin oil onto the surface of the salt water, this oil depth being at least 1/2" thick. Where paraffin oil is not available, boiled linseed oil, castor oil, cotton-seed and peanut oil have been used.

The wet condenser has the unusual property of preventing overheating of the glass jars since the convection of the salt water can take place. The liquid surface also keeps voltage uniform over the entire surface of the glass, preventing undue strain.

The level of the liquids should be at the same height both inside and outside the jar and ample air space left on top to prevent spark over.

One simple method of joining jars in parallel is as shown:

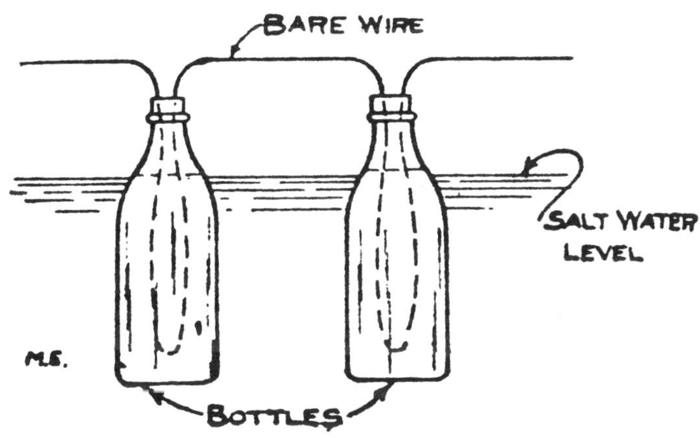

Circuit Interrupters: Spark Gaps

The main requirements of spark gaps are the ability to be adjustable with the circuit energized, freedom from oxidation of the gap electrode, and very rapid extinguishing of the arc. For low power Tesla coils, the series gap is sufficient. By using a series of gaps, a large cooling surface can be obtained:

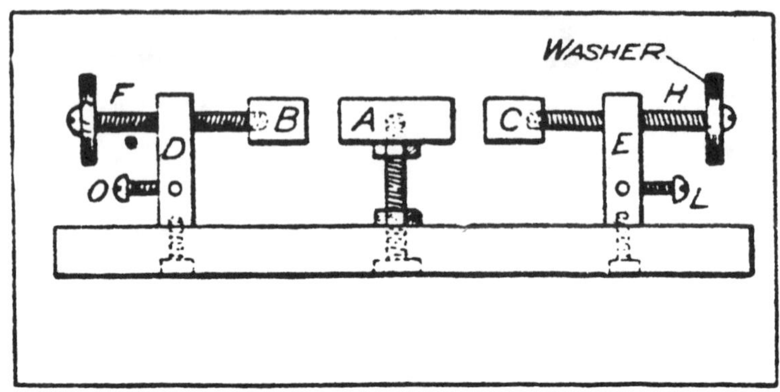

SERIES SPARK GAP

Fine thread screws give good adjustment. In all cases the preferred metals for electrodes are first zinc then brass, stainless steel or aluminum since these metals resist oxidation. Sometimes the insulated poles of a magnet placed close to and across the gaps are used to "blow out" the arc. A means for adjusting the gaps with power on is to add plastic tuner rods approximately 1/4" O.D. and 8" long to the electrode screws. Use extreme caution when adjusting the primary circuit.

Rotary or Mechanical Spark Gaps

Rotary spark gaps are essential for serious high power experiments and were standard equipment in Tesla's laboratory. One sample is given below:

ROTARY SPARK GAP

The main objection to the rotary type of spark gap is the fact that it is very bulky and takes up a great deal of valuable room on the operating table. In order to overcome this objection, the size is, in many cases, reduced to such an extent that the efficiency is greatly decreased. The accompanying drawing shows the construction of a type which has been designed to overcome this objection without loss in efficiency.

The rotary disk is made of hard rub-

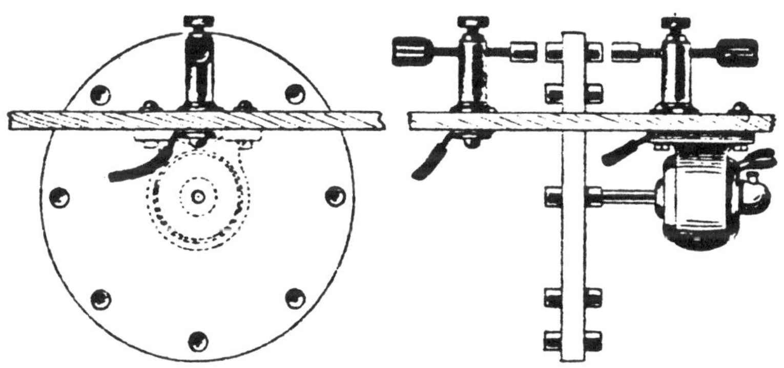

An Efficient Type of Rotary Gap.

ber and may be from 8 to 12 inches in diameter. Instead of having the electrodes on only one side of the disk or on the edge, as is the usual custom, it has a set on each side, the corresponding electrodes being connected by small threaded brass rods. The disk is connected to a motor, giving from 1,800 to 2,400 r. p. m., in the manner shown. The motor is fastened either under the operating table or back of the switchboard, so that the disk will project through a rectangular hole as shown. The stationary electrodes are placed on top of the table or on the front of the switchboard, and are insulated by means of hard rubber washers. They are provided with threaded shanks and hard rubber handles so that they may be adjusted to the most satisfactory distance.

The rotary gap gives a faster break, gives more breaks per second and where the motor speed is variable, more control over circuit interruptions. The spinning electrodes are also kept cool. From 8 to 16 electrodes on the disc are usually employed. Tesla's patented rotary gap gave almost 5,000 breaks per second.

For safety and noise reduction all forms of spark gaps should be enclosed in a fire resistant chamber. A window for observing the gap may consist of the rectangular glass filters used on arc welding helmets. When adjustments are properly made, the spark will be sharp and snappy, with a high-pitched tone, not quiet or flaming in appearance.

This is a description of type of spark gap which I believe to be of novel construction. It possesses several advantages, such as freedom from noise

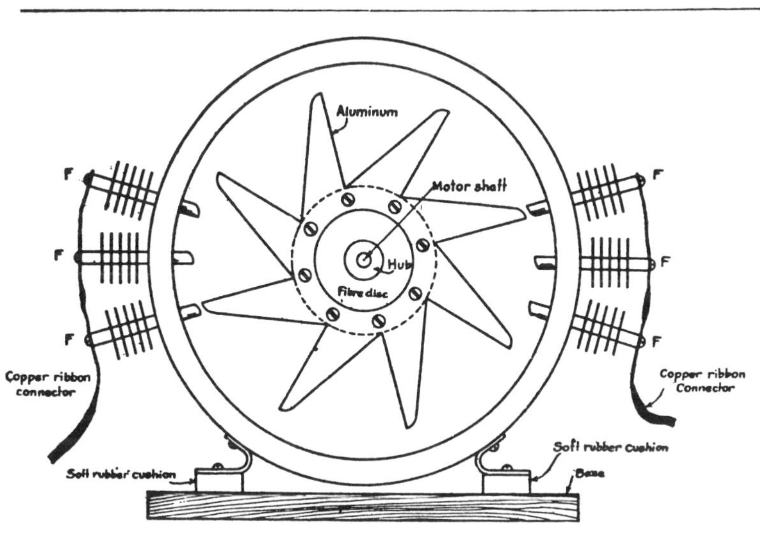

and airtightness; it gives quenched effects and, owing to the extreme lightness of the moving parts, allows quick starting and stopping with a very small motor. Referring to the diagram, Fig. 1, the rotating wheel is 7 inches in diameter and shaped somewhat after an 8-pointed star. It is cut from 1/16-inch aluminum and mounted on a 3-inch fibre insulating hub as shown. The wheel may be cut with ordinary tin shears and bored with a hand drill. It is then placed between flat slabs or boards under pressure, which should flatten out the aluminum, giving a true wheel.

The casing for enclosing the wheel should also be turned from fibre, although hard wood may be used at less expense if care is taken in the cutting, with the additional precaution of making the walls somewhat thicker. The inside diameter of this drum is 8 inches and the depth ¾ inch; the walls may be from ¼ inch to ½ inch. The cover is turned from ¼-inch fibre and screwed on tightly, making the casing airtight. If the hole for the motor shaft is made an easy fit and all work well done, the shaft will run in the centre of the hole without contact or vibration. The casing may be mounted as shown with brass angle pieces or in any convenient manner.

The fixed electrodes F are turned from brass and are supplied with cooling fins. They are made from ⅜-inch shanks with the outer ends tapped with an 8/32-inch thread to take a screw which is used to clamp the connecting ribbons from the condenser and inductance. The 3-pair arrangement of

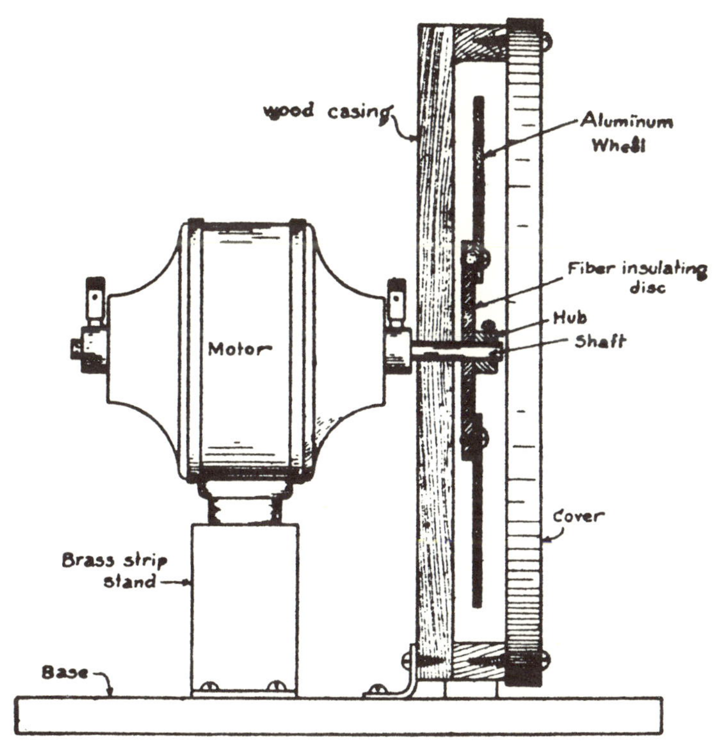

Fig. 2, First Prize Article

electrodes as shown, properly spaced, gives much better cooling than that afforded by the usual construction, for each electrode gets only every third spark during rotation. This gives, with an 8-tooth wheel, 24 sparks per revolution, producing a high tone with a low speed, whereas the ordinary gap would require a 24-tooth disc, a heavier wheel or as an alternative a small wheel and an extremely fast motor. My arrangement allows quick starting, which is further facilitated by using a 2-point switch in the motor circuit; that is to say, full power is thrown on the motor for quick acceleration, and when the switch is placed on the second point, a resistance coil is cut into the circuit.

The wheel should be carefully balanced on the hub before final mounting, shaving off the material on the heavy side. The motor I use is a high speed series Universal motor, which may be purchased at a cost of $5.00.

The speed regulator is made by winding No. 30 bare German silver wire on a 1-inch porcelain tube 10 inches in length. The wire may be easily spaced by winding two wires closely side by side, then fastening one and unwinding the other. The tube thus wound is suitably mounted and a slider attached to it, after the same manner employed in connection with tuning coils. If a very high voltage (15,000 or 20,000) is used, two or three plates of the ordinary quenched spark gap type should be used in series with the rotary, or if this is not done, the wheel may be made larger or two pairs of fixed electrodes may be used with a higher speed. The latter is probably advisable because in order to secure a good tone on any rotary gap, the electrodes must be set very close and the spark must not be allowed to jump two electrodes at one time, as would happen if the wheel were too small.

Fig. 2 is a side elevation of the gap, showing the mounting of the motor and the disc to the shaft. Fig. 3 is a detailed diagram of the base and the method of mounting the motor.

In regard to spark tones, it is my opinion that a low resonant tone is as pleasant as a high whistle, but I note

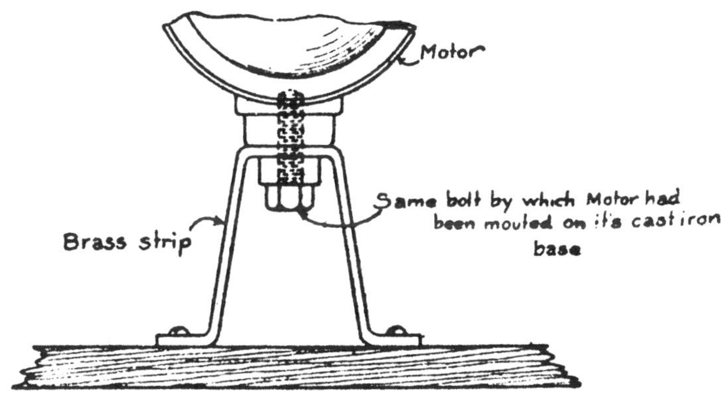

Fig. 3, First Prize Article

that the condenser capacity should be of small value if quenched effects are desired. I believe this gap will be a welcome relief to amateurs from the noise of the open rotary gap and from the ponderous flywheel type having a rotary disc with brass plugs.

Another method of regular interruption incorporates a tuning fork into the circuit. The sketch shows one used in an audio frequency oscillator.

Tuning Fork Oscillator.—An audio oscillator of fixed frequency utilizes a tuning fork set into vibration by magnetic action as illustrated in Fig. 7. A U-shaped fork is magnetized by the field coil connected to a battery. Battery current also flows through a microphone button, the metal of the tuning fork and the primary winding of the input transformer. Vibration of the fork alternately compresses and releases the carbon particles in the microphone button. The consequent change of resistance in the button varies the cur-

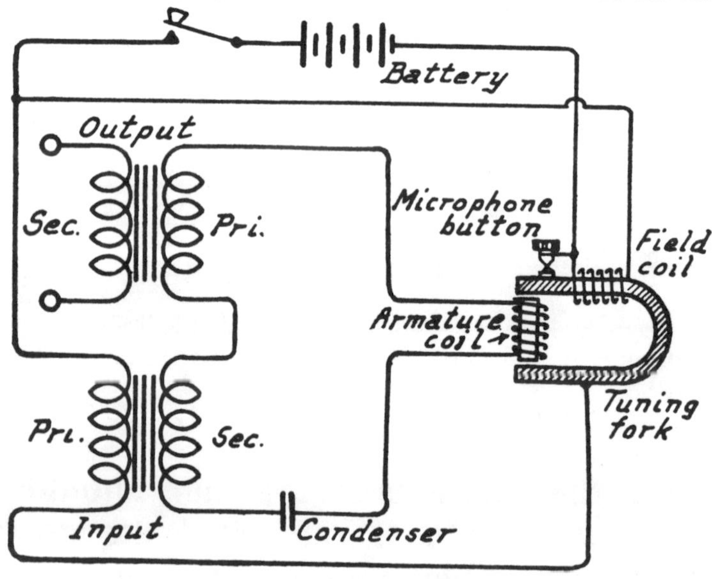

Fig. 7.—Tuning Fork Oscillator.

rent through the input transformer primary. A voltage is thus set up in the secondary of the input transformer and passes through the armature coil and the primary winding of the output transformer.

The alternating voltage and polarity of the armature causes it to attract and repel the tuning fork since the ends of the fork are the poles of a magnet produced by the field coil. Vibration of the fork at its natural frequency is maintained by this action. The circuit including the input transformer secondary, the output transformer primary and the armature coil is tuned to resonance at the fork's frequency by the condenser shown. The output of this oscillator is taken from the secondary terminals of the output transformer.

I have never found a Tesla coil circuit specifically using a tuning fork, but those who have used them for other oscillators state that distortion of frequency occurs when excessive amounts of current flow through the fork circuit so their use should be confined to small units.

Another type of circuit interrupter is the quenched gap. While it is the most efficient of the several types, its construction is somewhat difficult. It works better in low power circuits because overheating can cause warpage of the discs and burning of the sheet insulation employed. The mica ring material can be made from isinglass, used in older woodstoves. Besides mica, very thin (1/100 inch) surgical rubber sheeting available from dental supply houses may be used. It is called "dental dam". Not shown in the drawing is the frame which permits pressing the series of discs together to vary the setting.

TELEFUNKEN QUENCHED GAP

This is really a number of Lepel gaps connected in series. This arrangement can be substituted for the ordinary gap of a spark system. The disks are turned as shown from 3-16 or ¼ inch sheet brass to an outside diameter of 6½ or 7 inches and grooved 1 or 1½ inches in, so that the groove is about 3-8 of an inch wide at the face. Each plate is grooved on one side in this manner. See Fig. 67.

The mica rings used may be had at supply houses and

should not extend further in than 1-8 inch beyond the outside diameter of the groove, so that the inside circumference of the mica comes within ¼ inch of the in-

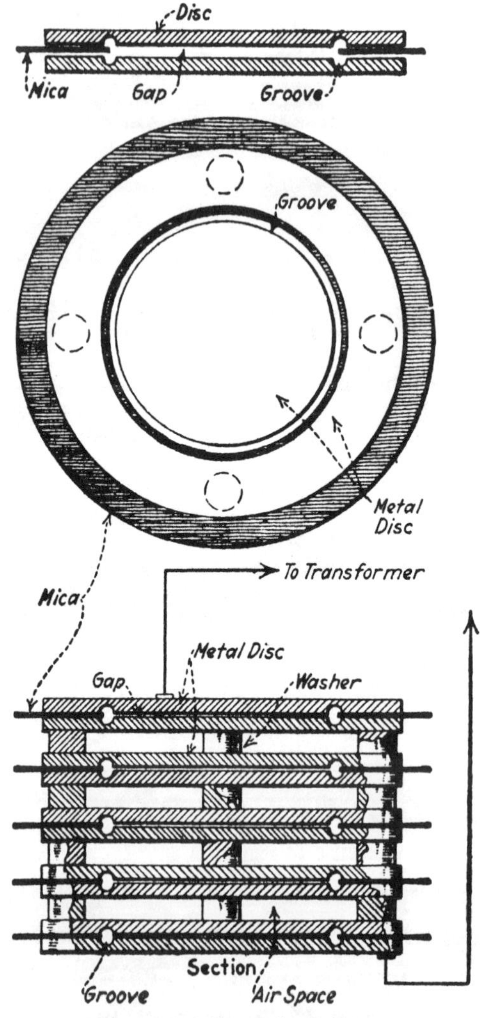

FIG. 67.—Quenched Gap.

side circumference of the groove. The groove is to prevent the spark from jumping to the mica as the latter becomes a conductor when heated by a high frequency discharge. The mica rings should not be more than .01

inch thick. The disks are assembled in pairs so that the grooved faces are next to each other, and washers are placed between the pairs so that the pairs are separated by a distance equal to the thickness of one of the plates. Thus if ¼ inch plates are used, the washers used should be ¼ inch thick. The assembled gap may be suitably mounted by using insulated supports, a sufficient number of pairs being used so that the combined length of the gaps is somewhat less than the length of a single gap, ordinarily used. When large power is used with this gap, it is well to have a small fan blow upon it to dissipate the heat which is generated.

THEORY AND ADVANTAGES OF THE QUENCHED SPARK

The gaps described are not difficult to construct and operate and are recommended to the readers. The discharge is practically noiseless, almost 60 per cent more efficient than a common gap, and produces nearly undamped waves. A high pitch note, which increases the effective transmission range, can also be had.

The operation of the quenched gap depends upon the fact that the spark quenches itself out after it has made a few oscillations, allowing the secondary oscillations to continue freely. This was illustrated in Figure 31. The primary circuit is thus opened so that it does not interfere with the secondary or aerial oscillations. As a result the unwelcome beats common to open spark systems are avoided. Returning to the parallel case of a gong, the quenched spark may be compared to a padded hammer, which after striking the gong (comparable to the antenna circuit in this case), a forceful blow, allows it to continue by itself with a clear, powerful vibration. The *short spark gap* when well cooled prevents the primary from *oscillating* by itself after the secondary circuit has been excited. That is, the spark is *active only long enough* to allow the secondary oscillations to reach a maximum, and the secondary oscillations are a maximum after the primary oscillations are reduced to a minimum. The number of primary oscillations necessary for this ideal operation is governed by the degree of coupling between the primary and secondary. It is desirable to use a close degree of coupling with the quenched spark for this reason. The energy ordinarily lost as heat in an ordinary spark gap is thus conserved and the wear on the primary apparatus is reduced. One of the chief causes of heat in the condensers and wear of the gap with an ordinary open gap is the useless continuance of the energy after the useful oscillations have been generated. The *quenched gap,* then, *prevents* undesirable oscillations from being set up in the primary by the reaction of the secondary, and makes the resulting radiations have a single wave length, for receiving purposes.

In constructing the quenched gap, it is essential that the electrodes be pressed with some force against each other.

Part Two - Frontiers for Experimenters

Wireless Power Transmission Experiments

As stated before, Tesla understood from his experiments at Colorado Springs in 1899 that the earth is highly charged naturally and that the globe behaves as though it were a part of a fixed spherical condenser. About this time, he received U.S. patents 645,576 and 649,621 which describe the "high potential magnifying transmitter".

These incorporated the flat pancake spiral form in the primary and secondary coils because this shape electrostatically coupled efficiently with oscillations in the earth and gave the purer sine wave needed for resonance.

Mr. Tesla's description of how the transmission of energy traveled is given here:

> • In describing the mode of transmission of his oscillating currents through the earth, Tesla claimed the path of the discharge was from his station directly through the center of the earth and in a straight line to the antipode, the return being by the same route, and that the current on this straight-line path traveled at its normal velocity—the speed of light. This flow, he declared, produced an accompanying surface flow of current, which was in step at the starting point and when they rejoined at the antipode; and this necessitated higher velocities in flowing over the surface of the earth. The surface velocities would be infinite at each of the antipodes, and would decrease rapidly until at the equatorial region of this axis it would travel at the normal velocity of the currents.

This description, of course, does not agree with modern views of electromagnetic radiation. However, Tesla understood better than his critics just how his devices worked and what principles were involved. Clearly this was something quite unlike mere Hertzian waves.

No. 645,576. Patented Mar. 20, 1900.

N. TESLA.

SYSTEM OF TRANSMISSION OF ELECTRICAL ENERGY.

(Application filed Sept. 2, 1897.)

(No Model.)

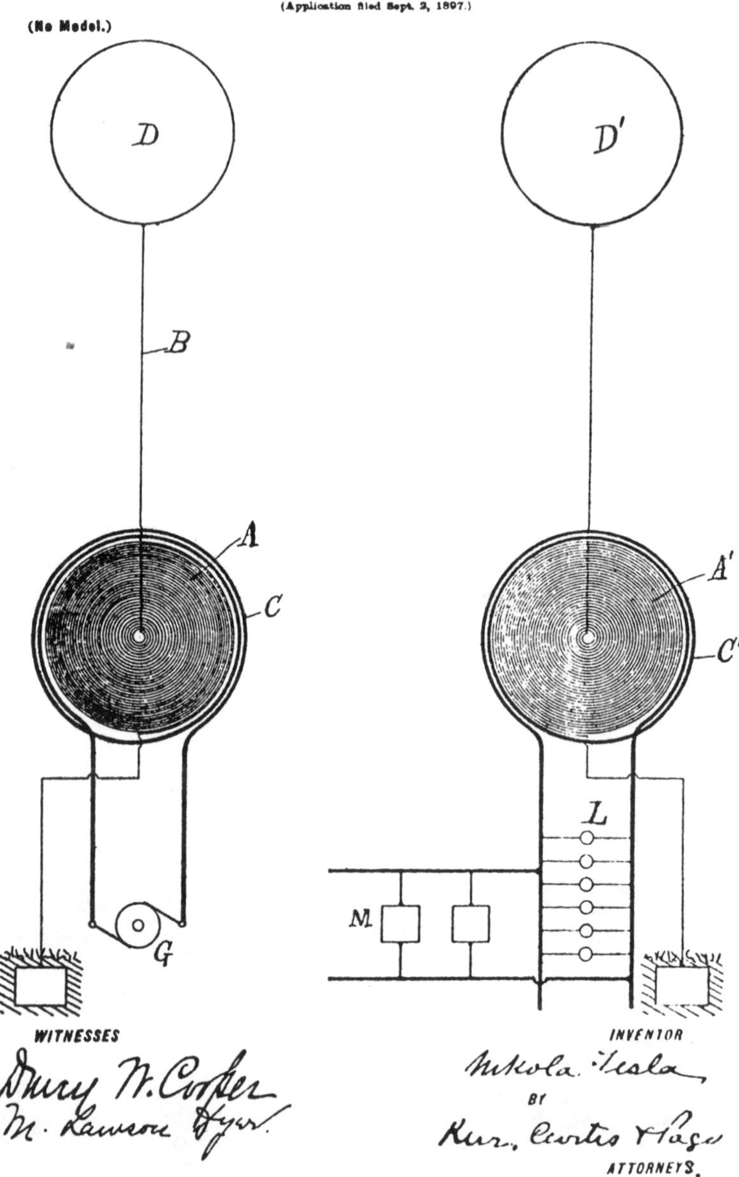

An often overlooked detail in Tesla's writings is mentioned in U.S. patent 645,576. On page 2, lines 131 to 133 we find (the secondary coil may be) "wound in spiral form either about a MAGNETIC CORE or not, as may be found necessary." Normally it is assumed that with high frequency Tesla circuits, air cores alone are suitable because of the problem of power losses in magnetic cores at high frequency. The first major break through in high frequency transformer efficiency was mentioned in 1915. The enclosed abstract describes an article (in German) by Ludwig Kuhn:

68. *Transformers for Undamped High-frequency Currents.* **L. Kühn.** (Helios, Nos. 33, 34, 35, 37, 1915. Jahrb. d. Drahtl. Tele. 11. pp. 133–194, Oct., 1916.)—The article furnishes a theory of transformers with closed magnetic circuits, together with particulars of their design and mechanical construction. The article is lengthy and not of a nature suitable for abstraction. The use of transformers with closed magnetic circuits is desirable on account of their small volume as compared with transformers with air cores that are commonly employed and on account of the avoidance by their use of large stray fields with their accompanying eddy-current losses. It is shown that the design of iron-core transformers is possible for wave-lengths up to 3 km. and even higher. The design is carried out on the basis of obtaining a minimum cost and minimum losses, and it is found that under these circumstances the greater part of the total losses lies in the copper. In order to avoid distortion of wave-form small induction densities—not above 600 lines per cm.^{2}—are used, and this precaution, together with the use of very thin laminations—0·03 to 0·05 mm. in thickness—of iron of low conductivity enables the hysteresis and eddy-current losses to be kept very small. The article concludes with the full calculation of a transformer to be used with an aerial of capacity 0·03μF, inductance 0·4μH, and radiation resistance 2·5 ohms, to emit a wave-length of 10 km. fed by a Goldschmidt high-frequency generator with a ground-frequency of $\frac{1}{4}$ of the transformer frequency. The total output is 100 kw. Formulæ are also developed for calculating the dimensions and prices of the tuning capacity and reactance. A. J. M.

Here we find an iron core transformer efficient at 100 kilocycles, the importance being that the iron laminations are of foil thickness. Much later, about 1934, W.J. Polydoroff described magnetic cores made from compressed iron powder which led to the ferrite cores often found in today's radio transformers.

The experimenter wishing to improve on transformer efficiencies should consider movable paramagnetic cores when working at the lower frequencies as well as the use of diamagnetic materials such as copper powder. This is another neglected area which Tesla possibly explored.

Additional advanced experimentation with wireless power transmission techniques was planned at Wardenclyffe, Long Island, New York. In a 1912 interview Tesla stated he intended to pay for these experiments by royalties realized from his newly patented bladeless steam turbine. When this did not materialize, this work had to abandoned in mid-stream.

Later, from 1925 to 1930, James Harris Rogers incorporated some of Tesla's wireless transmission discoveries in his underground communications system for which he received a patent. Even at this date many scientific "experts" in the field argued that buried antennas would not work efficiently. The Rogers' system became useful in submarine communications and was able to eliminate static interference. However, no credit was given to Tesla for his pioneering work. (See accompanying sketches of a typical underground antenna)

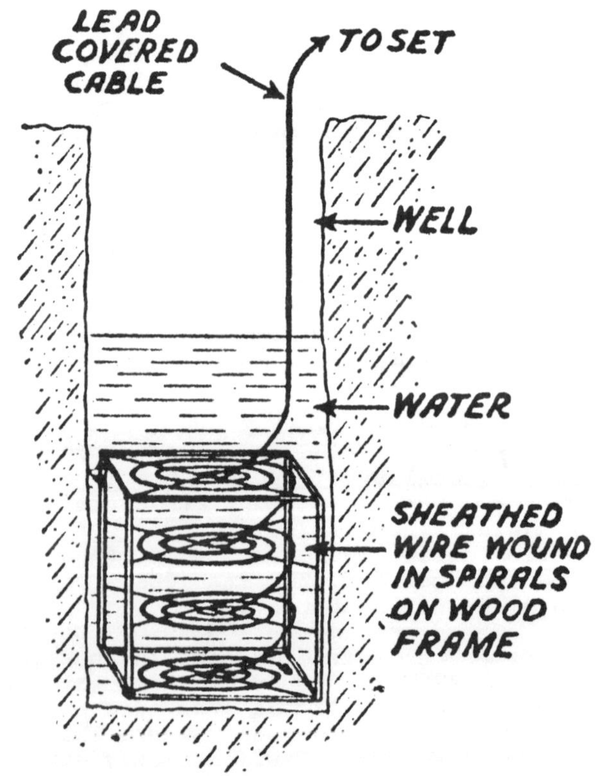

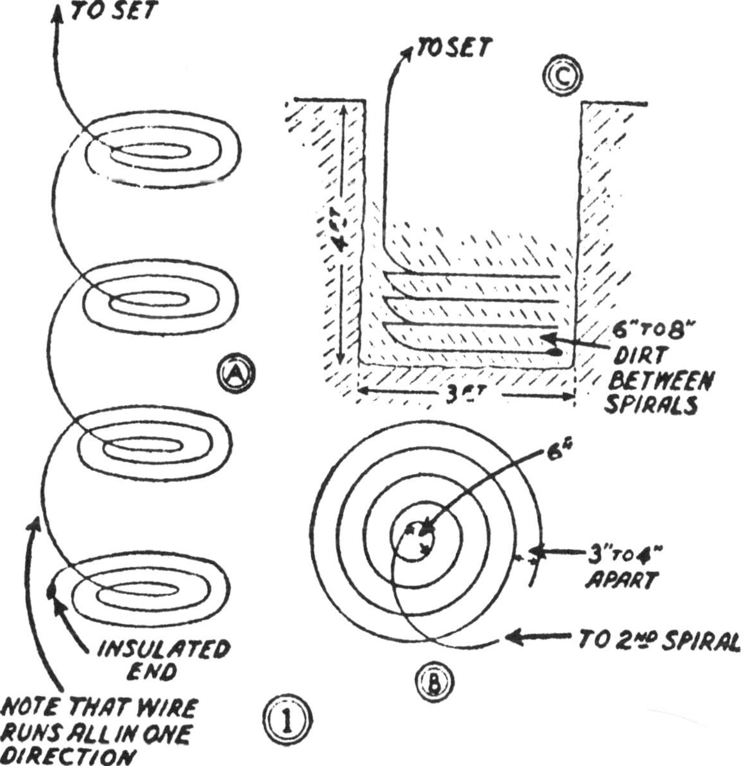

The illustration above shows how an underground antenna can be made with a minimum of labor—*Illustration Courtesy Cloverleaf Mfg. Co.*

Electrotherapeutics

In an interview given in 1899 Tesla recommended the use of Tesla coils for medical treatments of various kinds. At this time, the power supplies for electrotherapy treatment had been confined to high-voltage direct current electrostatic machines such as the Holtz or Wimshurst. Within a decade Tesla coils were being mainly manufactured for electrotherapeutic practitioners. Benefit was said to be in the form of localized deep heating effects of high-frequency currents (the beginning of diathermy treatment) for which Tesla received no credit.

Tesla did manufacture and did profit from medical coils for a time, but there were more interesting experiments to be done. He soon ended the enterprise.

Although Tesla did not experiment much along this line, others, especially in Europe did develop some unusual facets of high frequency electrotherapy to a remarkable degree. The most notable was the research of Gosset, Gutmann and Lakhovsky in Paris, France.

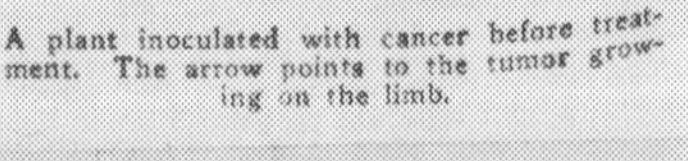

A plant inoculated with cancer before treatment. The arrow points to the tumor growing on the limb.

The plant after treatment. The arrow shows the part of limb where the tumor was.

The scientists innoculated plants with a bacteria that produced tumors. Later they exposed the plants to very high frequency radiations of low power intensity (150 million c.p.s.). The tumors were given eleven three-hour doses of radiation. Sixteen days after the first treatment they began to shrink and dry up. They were then easily brushed off the plant. Control experiments with the similar plants not treated showed tumors continuing to grow. From 1924 through 1941 Georges Lakhovsky expanded these experiments to successful results with cancer in animals and humans. His conclusive theory was that cells of the body resonate to particular frequencies of electromagnetic radiation. The RNA-DNA complex, in the form of a helix acts as a coil. Each cell responds as a tuned circuit, so Lakhovsky's oscillator produced broad band radiations or multiwave oscillations which were selectively picked up by the body's unhealthy cells. This raised the cell metabolic rate by electrolysis and restored reproductive capabilities and general vitality to the cell unit. Note that this method is in no way similar to the use of localized heating via the diathermy machine, which has a crystal-controlled monofrequency and higher power output.

WARNING: Since the Lakhovsky multiwave oscillator consumes about 40 watts of power, much of which is in the broadcast band, experiments must be performed in a metal shielded room to avoid F.C.C. violations. In 1934 Lakhovsky did receive U.S. patent #1,962,565 and later #2,351,055 which describes his multiwave oscillator. The circuit diagram is very similar to that for the Tesla coil. Besides Lakhovsky, Reginald Fessenden, the great pioneer in wireless communications, also claimed successful treatment of cancer in animals but details were not disclosed. See FESSENDEN-BUILDER OF TOMORROWS by Helen Fessenden (New York 1940), and THE SECRET OF LIFE by George Lakhovsky (1935), available from Health Research, Box 70, Mokelumne Hill CA 95245.

Photographs of experiments by Kristian Birkeland, a Norwegian scientist, conducted from 1908 to 1921 appeared in the January 8, 1921 issue of Scientific American magazine. These photos show the various patterns of light formed about an electrified copper sphere placed in an evacuated chamber. Specifically the aurora was simulated by placing a cylindrical electromagnet within the hollow sphere which resulted in the formation of magnetic poles. This simulated the magnetized earth in space. A negatively charged disc at the side of the box provided charged particles, and although these experiments were confined to high voltage direct current, much information could be learned about "cold light" and magnetic fields using the Tesla transformers. Notice that this work is similar to Tesla's lighting experiments with the single element spherical "carbon button" lamp.

In several of these experiments, power was supplied at 15,000 volts D.C. at 500 milliamps. Sphere diameter varied from 8 to 40 cm and the chamber was evacuated down to a few hundredths of a millimeter. See ON THE CAUSE OF MAGNETIC STORMS by Olaf Kristian Birkeland (London, New York 1908).

Electrohorticulture

This area of research was an attempt to increase growth rates and quality of vegetation by means of electricity. Experiments can be traced back to the year 1746 in Edinburgh, Scotland. In 1904, Professor Lemstrom published his work ELECTRICITY IN AGRICULTURE AND HORTICULTURE in which he describes the application of high voltage (60,000 to 100,000) direct current to a grid of overhead wires covering the garden area.

The Tesla coil was occasionally used as a power supply in experiments. Power requirements were normally only 75 to 100 watts per acre. Today, such experiments would need to be performed in shielded greenhouses or in isolated regions, of course. Yield increases for many types of crops reportedly averaged 10% to 50%.

The art was well-developed by experimenter Kenneth E. Golden who received U.S. patent number 1,952,588 in 1934 for his system. This area of work is quite similar to the discoveries made by Mr. Lakhovsky.

The response by plant life to low power oscillations raises the question of to what extent are very high voltage transmission lines involved in the formation of acid rain. The relationship could be indirect but still significant.

Theories on the formation of acid rain generally do not put emphasis on the presence of ozone, a very corrosive gas. The section discussing spark gaps indicates that only non-corrosive metals can be used where ozone is concentrated. The response by plants to very low power high voltage sources suggests that high levels of atmospheric ionization from electric power transmission lines, microwave radiation, radioactive gas emissions and wastes may be very deadly conditions to our forests. If it is not damaging to the vegetation itself, it could very well upset the bacterial level of balance.

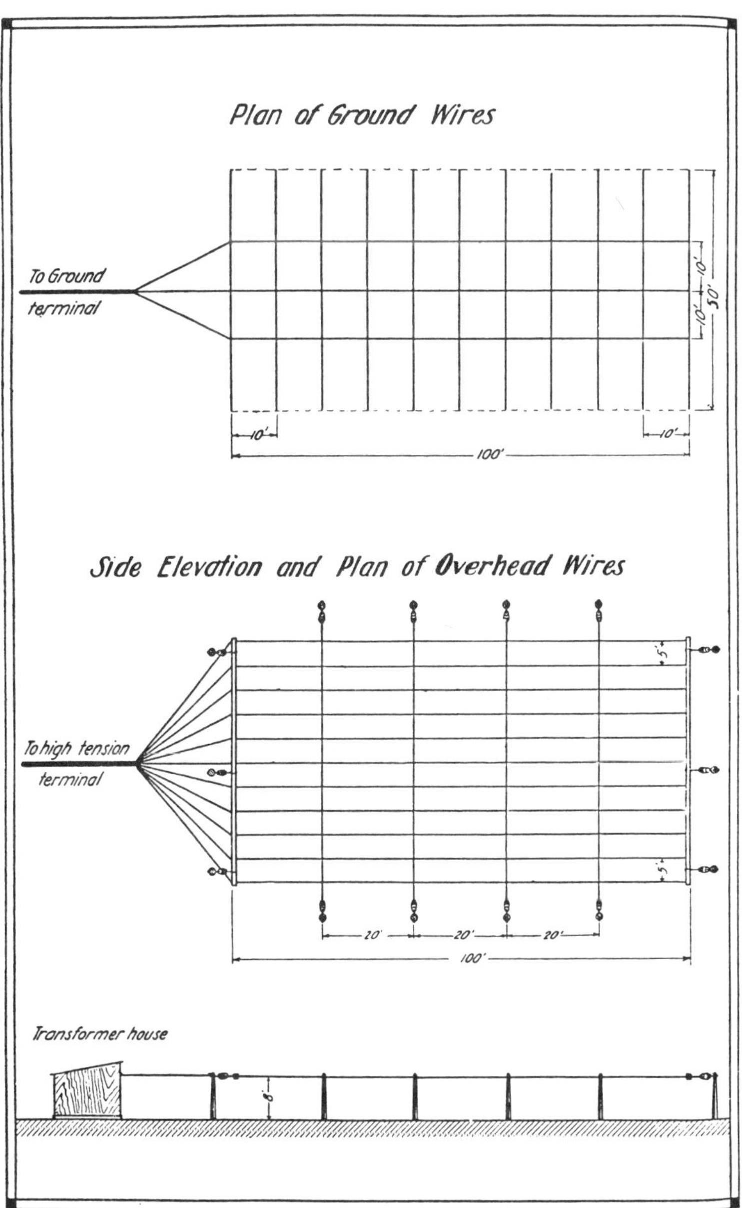

Plan of Ground Wires
To Ground terminal
10'
10'
50'
10'
10'
100'
Side Elevation and Plan of Overhead Wires
5'
To high tension terminal
5'
20'
20'
20'
100'
Transformer house
8'

Luminous Discharge Over Liquid Surfaces

Peculiar discharges over liquid surfaces such as water using high voltage battery cells were described by the experimenter Plante in his book STORAGE OF ELECTRICAL ENERGY in 1887.

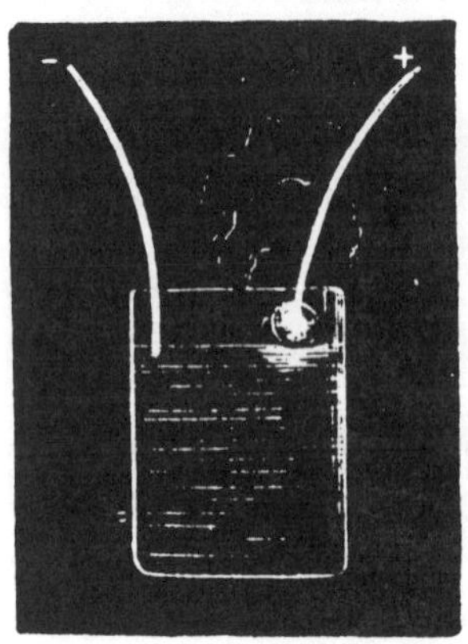

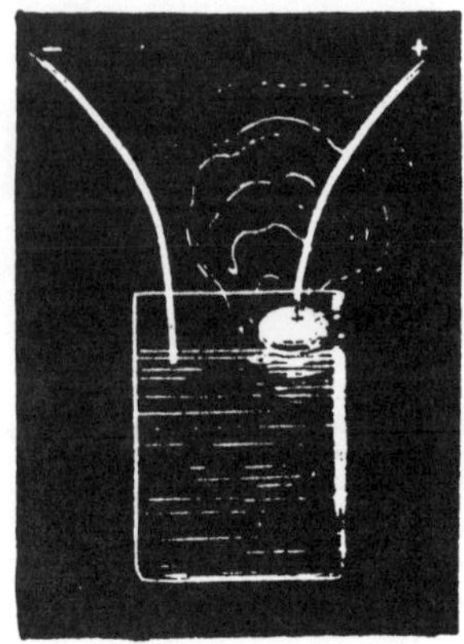

A negative electrode is slightly immersed in a salt water solution, and the positive wire is slowly brought close to the surface. When the arc begins, the positive wire is sightly withdrawn. A luminous liquid globule is formed up to about 1 cm. diamater. This globule acquires a rapid spiral motion and tends to flatten out. The power requirements were estimated to be 500 to 1000 volts D.C. at 250 milliamps.

Later in 1901, the scientist Hesehus extended Plante's work and used a high voltage transformer producing roughly 10,000 volts A.C. A copper plate separated from the water surface 2 to 4 cm. was used and balls of fire, rays, conical and ovoid luminous forms were produced.

Tesla, in his Colorado Springs experiments, was able to produce fire balls when his oscillator was working at a maximum power. Although he attempted to photograph them, results were poor because of the balls' momentary existance.

Recent attempts by Robert K. Golka at Wendover, Utah to reproduce Tesla's results have not been entirely successful. This appears partly due to his working in a location which does not have the proper environmental conditions, namely natural electrical activity and the presence of suspended moisture. Both of these conditions were present at the higher elevation in Colorado Springs. Most reports concerning "ball lightning" include the presence of semiconducting materials and an abundance of suspended moisture, a point lacking in modern theories on the formation of fireballs. In addition, winding forms of open wooden skeletons, to permit air flow over the coils, and the use of cotton-covered wire (which will absorb moisture) are details which could make crucial differences.

Very Low Frequency Oscillations and Their Effects

Experiments undertaken by the inventor Guglielmo Marconi in Italy during the spring of 1936 using low frequency waves revealed their extraordinary ability to penetrate metallic shielding. These waves could affect electrical devices, overload circuits and cause machines like generators, electric motors and automobiles to stall. Diesel engines, which do not depend on electrical ignition while running, were not affected. His experimental notes on the subject were not found after the war.

In 1941 a 15-year-old boy in Appleton, Wisconsin again discovered a particular frequency which shorted out all motor vehicles in a radius of 3 miles. The boy's oscillator was a shortwave set, however, we cannot rule out the existence of two similar circuits which were detuned slightly to produce beat frequencies at the low end of the spectrum. Further details on the boy experimenter are not available.

Reduction of Weight Using a Hertzian Wave Generator

In 1911, New York engineer and inventor Edward S. Farrow publicly described his work with a "condensing dynamo". When this electrical device was placed on weight scales and power applied, the dynamo and any small weight attached to it, proceeded to lose 1/6th of its weight. In the photo the total weight was 18 ounces before the power was on and it dropped down to 15 ounces while in operation.

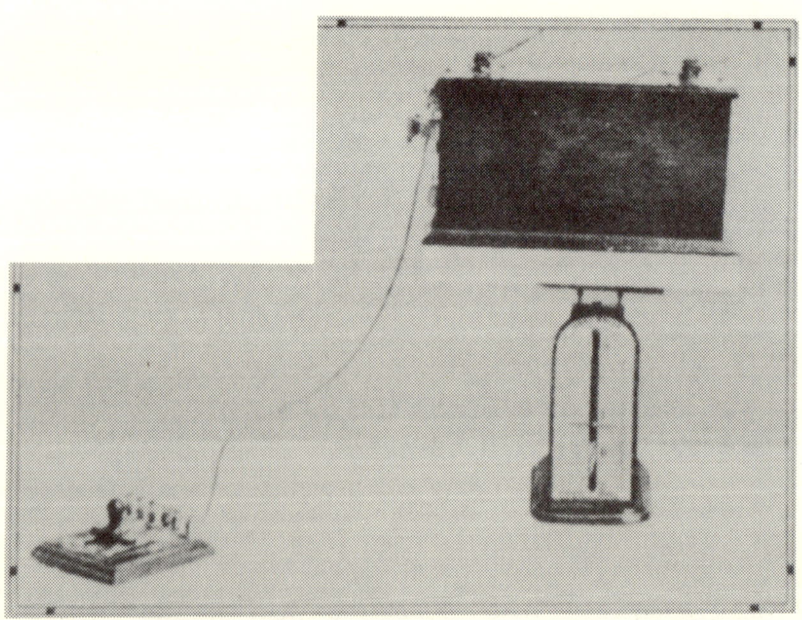

In Farrow's explanation, he said that the dynamo acted to "intensify the vertical component" of the Hertzian waves which it generated. This intensification produced buoyancy in any object to which the unit was attached. The unusual pattern of Hertzian waves fanned out in a thin horizontal plane of electromagnetic stress over a broad area.

The condensing dynamo employed either a horizontal row or a ring consisting of a series of interrupters or breaks (gaps) for producing this field. The ring of electrical breaks extended in a horizontal line around the perimeter of the device. Power and frequency of the oscillators were not given.

The buoyant effect is similar to that produced by floating a sewing needle on water. Although the steel has a much higher density than the water below it, the surface tension permits the linkup of many surrounding water molecules in a thin film or sheet. Similarly, the dynamo lifts against the strong gravitational field by reaction against the weak geomagnetic field. The interaction over a very wide area between this field and the Hertzian waves produces electrical buoyancy. No U.S. patent was received on the invention.

The reader can see from these descriptions many avenues open to serious experimenters using high frequency Tesla currents. Persons investigating wireless power transmission should bear in mind that the technology is not without its drawbacks. Impressing electrical oscillations on the globe would do serious environmental damage, as was indicated in some side effects of Tesla's short experiments in 1899.

In addition to environmental damage, such a method of electrical power transmission does not address the inherent weaknesses of a highly centralized and monopolized power base, such as exists today. A more plausible solution is the small-scale self-sustaining electrical generator, the space energy receiver.

As more information is discovered on the many varied experiments performed by Nikola Tesla, it becomes clear that he discovered a number of basic principles in electrical science which have yet to be recognized in orthodox circles. His work naturally led him to embrace peculiar views which resulted in difficulties both financially and professionally. A simple scan of the indices of classical textbooks on electrical engineering proves just how ignored his work was for many years. The recent resurgence in interest in this great pioneer does justice to all that he labored to bestow on the world society.

Sources of Information

General Reading:

NIKOLA TESLA - LECTURES PATENTS, ARTICLES by Tesla Museum Belgrad Yugoslavia
(1959) (Available through Lindsay Publications)

PRODIGAL GENIUS - THE LIFE OF NIKOLA TESLA by John J. O'Neill, published by I. Washburn Inc. 1944 (Available through Panther Books Ltd, London)

HIGH FREQUENCY APPARATUS, ITS CONSTRUCTION & PRACTICAL APPLICATION by Thomas Stanley Curtis published about 1916 by Everyday Mechanics Co. NY,NY.

THE TESLA EXPERIMENT - LIGHTNING AND EARTH ELECTRICAL RESONANCE by Charles A. Yost (Available from Tesla Books Co., 1580 Magnolia Ave, Millbrae CA 94030)

BIBLIOGRAPHY

SCIENCE & INVENTION Feb, Mar 1927 & Sep 1926
WIRELESS AGE June 1919
MODERN ELECTRICIAN 1915
POPULAR ELECTRICITY AND MODERN MECHANICS 1914
PHYSICAL REVIEW Vol 35 1930 (5,000,000 volt machine)
SCIENTIFIC AMERICAN V88 1903
POPULAR MECHANICS 1922
JOURNAL OF THE OPTICAL SOCIETY OF AMERICA & REVIEW OF SCIENTIFIC INSTRUMENTS Vol 18 1929 (Leyden Jar Photo)
RADIO NEWS 1925 (plant photos)
TECHNICAL WORLD magazine 1911
EXPERIMENTAL TREATISE ON PHYSICS by E Atkinson 1906 by William Wood & Co, NY
INDUCTION COIL DESIGN by M. A. Codd (1920) E. Spon Ltd, London
A PRACTICAL ELEMENTARY MANUAL OF MAGNETISM AND ELECTRICITY by Andrew Jamieson 1909 London
PRODIGAL GENIUS by John J O'Neill
EXPERIMENTAL WIRELESS STATIONS by Philip Edelman (1922) Henley Publishing Co NY

Since most of the material in this book was gathered over several years without there ever being any intention of publishing, detailed source notes were not kept. There may be sources other than those listed above that have been unintentionally omitted.

Equipment suppliers:

High voltage step up transformers, motors, speed controls:

Surplus Center
Box 82209
Lincoln NE 68501

Jerryco
601 Linden Pl
Evanston IL 60202

Insulating oils and test tubes available from:

Hagenow Laboratories, Inc.
1302 Washington St.
Manitowoc WI 54220

Other sources for small step up transformers include furnace contractors for used oil burner transformers and neon sign shops for sign tranformers. Electrodes for gaps may be in the form of solid brass drawer pulls or stainless steel ball bearings. Aluminum blower blades can be used in rotary spark gaps if insulated from the motor shaft. (Suppliers of paraffin oils are found in the THOMAS REGISTER under the subject of Oils, Paraffin, but they usually don't sell small quantities. Hagenow sells paraffin oil by the pound.)

Standard Copper Wire Table (reproduced from ARRL's 1934 Radio Amateur's Handbook)

Gauge No. B. & S.	Diam. in Mils[1]	Circular Mil Area	Turns per Linear Inch[2] Enamel	S.S.C.	D.S.C. or S.C.C.	D.C.C.	Turns per Square Inch[2] S.C.C.	Enamel S.C.C.	D.C.C.	Feet per Lb. Bare	D.C.C.	Ohms per 1000 ft. 250 C.	Current-Carrying Capacity at 1500 C.M. per Amp.[3]	Diam. in mm.	Nearest British S.W.G. No.
1	289.3	82690	—	—	—	—	—	—	—	3.947	—	.1264	55.7	7.348	1
2	257.6	66370	—	—	—	—	—	—	—	4.977	—	.1593	44.1	6.544	3
3	229.4	52640	—	—	—	—	—	—	—	6.276	—	.2009	35.0	5.827	4
4	204.3	41740	—	—	—	—	—	—	—	7.914	—	.2533	27.7	5.189	5
5	181.9	33100	—	—	—	—	—	—	—	9.980	—	.3195	22.0	4.621	7
6	162.0	26250	—	—	—	—	—	—	—	12.58	—	.4028	17.5	4.115	8
7	144.3	20820	—	—	—	—	—	—	—	15.87	—	.5080	13.8	3.665	9
8	128.5	16510	7.6	—	7.4	7.1	—	—	—	20.01	19.6	.6405	11.0	3.264	10
9	114.4	13090	8.6	—	8.2	7.8	—	—	—	25.23	24.6	.8077	8.7	2.906	11
10	101.9	10380	9.6	—	9.3	8.9	87.5	84.8	80.0	31.82	30.9	1.018	6.9	2.588	12
11	90.74	8234	10.7	—	10.3	9.8	110	105	97.5	40.12	38.8	1.284	5.5	2.305	13
12	80.81	6530	12.0	—	11.5	10.9	136	131	121	50.59	48.9	1.619	4.4	2.053	14
13	71.96	5178	13.5	—	12.8	12.0	170	162	150	63.80	61.5	2.042	3.5	1.828	15
14	64.08	4107	15.0	—	14.2	13.3	211	198	183	80.44	77.3	2.575	2.7	1.628	16
15	57.07	3257	16.8	—	15.8	14.7	262	250	223	101.4	97.3	3.247	2.2	1.450	17
16	50.82	2583	18.9	18.9	17.9	16.4	321	306	271	127.9	119	4.094	1.7	1.291	18
17	45.26	2048	21.2	21.2	19.9	18.1	397	372	329	161.3	150	5.163	1.3	1.150	18
18	40.30	1624	23.6	23.6	22.0	19.8	493	454	399	203.4	188	6.510	1.1	1.024	19
19	35.89	1288	26.4	26.4	24.4	21.8	592	553	479	256.5	237	8.210	.86	.9116	20
20	31.96	1022	29.4	29.4	27.0	23.8	775	725	625	323.4	298	10.35	.68	.8118	21
21	28.46	810.1	33.1	32.7	29.8	26.0	940	895	754	407.8	370	13.05	.54	.7230	22
22	25.35	642.4	37.0	36.5	34.1	30.0	1150	1070	910	514.2	461	16.46	.43	.6438	23
23	22.57	509.5	41.3	40.6	37.6	31.6	1400	1300	1080	648.4	584	20.76	.34	.5733	24
24	20.10	404.0	46.3	45.3	41.5	35.6	1700	1570	1260	817.7	745	26.17	.27	.5106	25
25	17.90	320.4	51.7	50.4	45.6	38.6	2060	1910	1510	1031	903	33.00	.21	.4547	26
26	15.94	254.1	58.0	55.6	50.2	41.8	2500	2300	1750	1300	1118	41.62	.17	.4049	27
27	14.20	201.5	64.9	61.5	55.0	45.0	3030	2780	2020	1639	1422	52.48	.13	.3606	29
28	12.64	159.8	72.7	68.6	60.2	48.5	3670	3350	2310	2067	1759	66.17	.11	.3211	30
29	11.26	126.7	81.6	74.8	65.4	51.8	4300	3900	2700	2607	2207	83.44	.084	.2859	31
30	10.03	100.5	90.5	83.3	71.5	55.5	5040	4660	3020	3287	2534	105.2	.067	.2546	33
31	8.928	79.70	101.	92.0	77.5	59.2	5920	5280	—	4145	2768	132.7	.053	.2268	34
32	7.950	63.21	113.	101.	83.6	62.6	7060	6250	—	5227	3137	167.3	.042	.2019	36
33	7.080	50.13	127.	110.	90.3	66.3	8120	7360	—	6591	4697	211.0	.033	.1798	37
34	6.305	39.75	143.	120.	97.0	70.0	9600	8310	—	8310	6168	266.0	.026	.1601	38
35	5.615	31.52	158.	132.	104.	73.5	10900	8700	—	10480	6737	335.0	.021	.1426	38–39
36	5.000	25.00	175.	143.	111.	77.0	12200	10700	—	13210	7877	423.0	.017	.1270	39–40
37	4.453	19.83	198.	154.	118.	80.3	—	—	—	16660	9309	533.4	.013	.1131	41
38	3.965	15.72	224.	166.	126.	83.6	—	—	—	21010	10666	672.6	.010	.1007	42
39	3.531	12.47	248.	181.	133.	86.6	—	—	—	26500	11907	848.1	.008	.0897	43
40	3.145	9.88	282.	194.	140.	89.7	—	—	—	33410	14222	1069	.006	.0799	44

[1] A mil is 1/1000 (one thousandth) of an inch.
[2] The figures given are approximate only, since the thickness of the insulation varies with different manufacturers.
[3] The current-carrying capacity at 1000 C.M. per ampere is equal to the circular-mil area (Column 3) divided by 1000